Agriculture and Resilience in Australia's North

Keith Noble • Tania Dennis • Sarah Larkins

Agriculture and Resilience in Australia's North

A Lived Experience

Keith Noble
James Cook University
Townsville, QLD, Australia

Tania Dennis
Insideout Architects
Townsville, QLD, Australia

Sarah Larkins
College of Medicine and Dentistry
James Cook University
Townsville, QLD, Australia

ISBN 978-981-13-8354-0 ISBN 978-981-13-8355-7 (eBook)
https://doi.org/10.1007/978-981-13-8355-7

This Springer imprint is published by the registered company Springer Nature Singapore Pte Ltd.
The registered company address is: 152 Beach Road, #21-01/04 Gateway East, Singapore 189721, Singapore

Introduction

Fly away little bluebird, I guess you'll have to find another farm. (Mark Knopfler[1])

For the reader to understand the orientation and motivation, and because the present is so heavily influenced by and cannot be separated from the past,[2] the following is some primary author context:

Early in my third year of school, we young students had to tell the class about ourselves. I told my classmates I wouldn't be around for long, as my dad was going to buy a farm and we would be moving. Nine years later, I completed high school just up the road from that primary school – still in Brisbane. My father had never actually said we were moving, but he did talk a lot about buying a farm, we did look at a few, and we ran cattle on agisted land outside Brisbane, so my young mind had just joined the dots. The concept obviously had appeal and stuck.

November 2000: 4 days drive from the Gibson Desert of Western Australia to Far North Queensland in a truck containing our worldly possessions and three red dogs sharing the cab, from one of Australia's driest regions to Tully, a *Pretty Wet Place*[3] in the Queensland Wet Tropics, where they measure rainfall in fathoms,[4] we were moving to our farm – a 31-ha established tropical fruit and taro property rising up the wet tropical slopes of the northeast of Tully – and the wet season had arrived early.

Thirty-six years after first talking about it, I moved to a farm – our farm. Though I won't pretend this had been an unwavering course through life, when there was a choice to be made, generally, the path that headed towards a farming future was the

[1] From the song *Bluebird*. On *Privateering* [LP]. Mercury Records Ltd.

[2] Bourdieu, P. (1984). *Distinction: a social critique of the judgement of taste* (R. Nice, Trans.). London: Routledge & Kegan Paul; and Berry, T. (2015). *The Dream of the Earth*. Berkeley: Counterpoint Press.

[3] The slogan on the Tully 'Gumboot', a 7.9 m rubber boot replica at the entrance to Tully that indicates the height of Tully's record annual rainfall, recorded in 1950.

[4] A unit of length in the old imperial and the U.S. customary systems equal to 6 feet or 1.8288 metres, used for measuring the depth of water. Used in a humorous sense by Tully residents when describing their rainfall.

one chosen: agricultural science at the university, first job at an agricultural service town and then steadily working west. I enjoyed a range of jobs in agriculture and natural resource management for the state and federal governments and discovered the joy of being part of small regional communities where the butcher sold raffle tickets at the RSL Club on Friday nights and ran pony club on Sunday, where everyone knew everything about everyone and what they didn't know they made up; but generally, with a care and consideration for other people that once experienced, I valued.

From Gatton to Longreach to Broken Hill (where I met my wife), to Charters Towers to Townsville (where I married her), then Alice Springs and the Warburton Ranges (a Ngaanyatjarra Aboriginal community in Western Australia) and all the small places in between, it was the same: people looked out for people and respected and appreciated the contributions made by others. The pinnacle, in my mind, was always those people who worked directly with the land – they had a pride and confidence that I admired and aspired to be part of.

So, in the early hours of Monday, 20 March 2006, when Cyclone Larry blew out the 600 new fruit trees we'd planted, it was only natural our first thought was to rebuild our farm and the emerging tropical fruit industry we had become part of. We'd had our one-in-a-hundred-year storm, so we'd be right now!

Third of February 2011: Severe Tropical Cyclone Yasi, category five, crossed the Far North Queensland coast near Mission Beach, between Cairns and Townsville, bringing peak wind gusts estimated at 285 kilometres per hour. The eye went over the top of our farm, giving 40 minutes of deceptive calm before the winds snapped back from the opposite direction – demolishing everything that the first onslaught had weakened. Not only were the 800 trees we'd replanted after Larry (and picked our first fruit from the previous year) gone, along with our new machinery shed and a good part of our roof, that night, the tropical fruit industry ceased to exist as an industry, with survivors slipping back 20 years to a state of disconnected individual growers.

Those trees weren't getting planted a third time, but I didn't know what to do next – the farming picture in my head seemed a permanent fixture. Certainly, the next slide wasn't queued and ready to view. This book is a direct outcome of that shattered dream. It is in part me, making sense of what happened – moving beyond feeling a victim of circumstances and understanding and accepting *life is what happens to you when you're busy making plans.*[5] It is also, in part, an attempt to understand why I wanted to 'go farming' in the first place when my work experience had shown it was 'a hard game'. But more importantly, this book aspires to provide an understanding of the operational context of Northern Australian farmers at a time when substantial industry expansion is being actively promoted.

[5] John Lennon (1980). *Beautiful Boy*. On Double Fantasy [LP]. The Hit Factory, New York.

Contents

About the Authors

Keith Noble trained as an Agricultural Extension Officer and worked in agriculture and natural resource management throughout Australia before farming in his own right in Queensland's Wet Tropics. With cross-cultural communication skills and expertise, particularly with regional Australians and their communities, Keith is able to solve problems, identify opportunities, and provide solutions that build on the inherent strengths of people and their place. Keith chairs Terrain NRM, the Regional Natural Resource Management body for Queensland's Wet Tropics bioregion.

Tania Dennis is an architect from the top end of Australia, who has designed an impressive series of places and buildings that allow people access to healthy living. Winner of numerous architecture awards, including the National Award for Small Project Architecture and Commendation for Sustainable Architecture, Tania's projects respond to and offer positive built spaces that influence how communities function and are perceived. By working with local people, makers and artists, and using skills applied through local culture to architecture, interior and urban design, Tania's work offers intricate and insightful interpretations of place.

Sarah Larkins is an academic general practitioner and Associate Dean of Research at the College of Medicine and Dentistry, James Cook University. Sarah has particular skills and experience in Aboriginal and Torres Strait Islander health research and health services as well as workforce research, and is an internationally recognised expert on social accountability in health professional education. Sarah is also Co-Director of the Anton Breinl Research Centre for Health Systems Strengthening, a centre of the Australian Institute of Tropical Health and Medicine. Sarah's particular focus is on collaborating to improve equity in health care services for underserved populations, particularly rural, remote, Indigenous and tropical populations, and on training a health workforce with appropriate knowledge, attitudes and skills for this purpose.

List of Figures

List of Tables

Part I
North Australian Theory and Realities

Chapter 1
Agriculture as a Human Endeavour

The question of questions for mankind – the problem which underlies all others, and is more deeply interesting than any other – is the ascertainment of the place which Man occupies in nature and of his relations to the universe of things.
Huxley (1863)

This book is set in and specifically relates to Northern Australia, but it is not a story of Northern Australia. It is a story of the people who both live and farm there, and their resilience strategies to cope with the region's inherent social and physical uncertainties. A dominant agricultural paradigm operates in Australia (and many other parts of the developed world) that is in keeping with Descartes insistence that *facts and values must be kept in separate worlds*, and so agricultural discussions are often in the context of facts, figures, yields, production schedules, contribution to GDP, seasonal forecasts and return on investment. Talk of humanity in agriculture is often missing or discounted to a subheading, particularly when new development is the focus (Fig. 1.1).

Agriculture however is a human endeavour. Its successful practice is the foundation of life, society and civilisation, especially when considered as a broader concept than that conjured by twentieth-century broadacre and intensive livestock operations. Most farmers are driven by a love of the land as much as a desire for profit, and there is an emerging body of work illustrative of the growing international awareness of the value of our Earth and its environment to a healthy and productive society and of our collective and individual responsibility to care for the Earth, as demonstrated in Australia by people such as Bill Gammage (2011) in *The Biggest Estate on Earth*, Bruce Pascoe (2014) in *Dark Emu*, and Charles Massy's (2017) *Cry of the Reed Warbler*. As with any human endeavour, the success of agriculture is also entwined with the complexities that allow people to meet their needs and are critical to their wellbeing: where and how they live and work, their educa-

Keith Noble has contributed more to this chapter.

K. Noble et al., *Agriculture and Resilience in Australia's North*,
https://doi.org/10.1007/978-981-13-8355-7_1

Fig. 1.1 A popular Australian perception about northern development. (Photo used with the kind permission of Mark Thomson, Institute of Backyard Studies www.ibys.org)

tion, opportunities for socialisation and healthcare support, along with relation and access to sites of spiritual, cultural and emotional value.

In recognition of and sympathetic to this human dimension of agriculture, a series of vignettes[1] are included throughout this book. While these vignettes illustrate specific points, they are equally intended to remind the reader that agriculture is a human endeavour, that farmers are people and that everyone's situation is unique and their story personal.

Robert Christison is the subject of the first vignette. He came from Scotland to Australia in its early colonial era with nothing, settled in the north, and left a wealthy man – but not by choice. He was torn between the love of his adopted country, for which he was a great champion, and his family, who suffered the trials of climate and deprivations of isolation, along with limited access to health and education services. And so it continues today: while many women effectively operate in and champion Northern Australia, it remains predominately a culture of masculine challenge necessitating sacrifice by other family members which, until addressed, will continue to limit development of a diverse and stable resident society.

But there is much more in the Christison story: in an era renowned for the exploitation of and cruelty towards Australia's First People and Traditional Owners, Christison championed their cause. He looked globally for new markets and innovation, and demonstrated an ethos of environmental awareness and sustainable land

[1]A **vignette** (written on a vine leaf) in literature means a short impressionistic scene which gives particular insight into a character, idea or setting. Here, the vignettes are differentiated from the main body of text by different font and line spacing.

management. In fact, this one man's life could be an allegory for a vision that continues to manifest right through to the present-day aspirations described in the 2015 *Our North, Our Future: White Paper on Developing Northern Australia* (Australian Government 2015).

Christison's story reveals how from the time of very early European settlement there has been an agricultural development vision for Northern Australia that many have strived to achieve, while the perceived obstacles continue to be relatively the same: distance, disaster, an unsympathetic southern government. Just because Christison did not achieve his vision in its entirety does not mean he was unsuccessful. He was very successful as an individual, and those that came after built on his success. But that is not the story of Northern Australia – it is not the story of Australia – it is the story of people and society. Whether Christison's or any other's vision of North Australian development is achieved is not a question of success or failure, it is a matter of whether an agreed collective vision for the North can be established, and decision-making processes agreed to that work towards achieving it.

Vignette 1.1: Christison of Lammermoor

Robert Christison (1837–1915) left Scotland aged 15 on a borrowed fare for Australia's goldfields. Arriving in Melbourne without friends or money, he found sheep work on the Werribee and discovered a sense of fellowship and pride in the bush. From sheep he turned to horsebreaking, survived attack by bushrangers, quickly abjured the lottery of mining and, after taking lessons in navigation, struck out in 1863 for new country which he might farm. Travelling north from Bowen (Queensland), he chanced upon the explorer William Landsborough, whose account of his search for Burke and Wills Christison knew by heart. Following a route described by Landsborough, Christison came to the Great Divide's western watershed and air that was 'lighter and drier … of diamond brightness' and, in his mind, saw the Lammermoor Hills in Scotland. This was to be his home.

He built sheep yards, horse yards and then a house, while determinedly making peace with the resident Dalleburra Aboriginals – 'country belonging to you: sheep belonging to me'. Joined by his two brothers, they weathered the depression of 1866–1869 when wool sold for seven pence a pound, the local bank failed and nearly a thousand runs were abandoned in the unsettled districts. In 1870 a cyclone destroyed Townsville and flooded the west – landholders who didn't know they'd built by a watercourse spent weeks on their roof watching trees and hard work flow by. Floods were followed by financial crisis, so Christison set off on 9-month journey walking 7000 sheep to Adelaide. Discovery of gold at Charters Towers in 1872 brought hope to the region; Christison bred Hereford cattle and prospered.

But brother Willie drowned crossing the flooded Burdekin River, which turned Robert's mind to family; so, in 1877 he returned to Edinburgh, gifted a third of his lands to his sister and her husband, and fell in love. Though his wife declared Lammermoor heaven, she succumbed to malaria, leaving Christison bereft and melancholy. Queensland's trade in tinned meat collapsed when the USA began shipping fresh meat in ice to Britain.

Challenge drove innovation, and Thomas Mort established the world's first freezing works in Sydney, declaring 'the time has arrived … when the over-abundance of one country will make up for the deficiency of another'. Christison again visited Scotland, remarried and in his enthusiasm established The Australian Company which built freezing works at Bowen. On the evening of 29 January 1884, beef was being frozen to ship next morning on the *Fiado*, when an ominous silence proclaimed a change – a cyclone. By noon the next day, the wreck was complete: the *Fiado* beached, and Bowen in ruins. Christison refused to deplore the loss 'for we shall yet see factories studded all over Australia'.

But drought followed, and the death of a son; a change of government raised rents and resumed leaseholds with no entitlement to compensation, resulting in 'stagnation and widespread misery … more ruinous than drought'; and the closing of all freezing works. Christison's brother died suddenly, and his wife declared she would return to London for the sake of the children.

Visiting his family, Christison called at Pasteur's Paris laboratory and learnt of a stream of artesian water obtained from 1800 ft, and his active mind turned to the spectacular floods that disappeared into the porous desert sandstone and rarely reached western waterholes. He ordered a boring plant in Sheffield which proved inadequate for Australia's hard rock, but a subsequent Canadian plant struck water.

In 1898, 46 years after arriving in Australia on a borrowed passage, Lammermoor had 500 Arabian horses, 40,000 cattle and a stud herd second to none, with every slab of the house and post in the yards split by Christison himself. The 1899 drought proved the wisdom of Christison's light stocking and network of sub-artesian wells; but as the drought continued through 1903, the effort of maintaining stock and pumps and an increasing overdraft took their toll on his heath, and he pined for his family. So, when rain came he decided he must sell to provide for them but would not discuss anything 'till [the Aboriginal's] right to remain on the station as their home is settled'. Robert returned to Scotland and is buried at the foot of the Lammermoors.

This vignette is extracted from M. M. Bennett's (1927) *Christison of Lammermoor*, and the story does not end here. Mary Montgomerie Bennett was Robert Christison's daughter and biographer, who materially benefitted from her father's fortune and, later in life, returned to Australia from England to became an activist for indigenous justice at a time when Aboriginal Australians had their citizens' rights curtailed by repressive state laws, and six decades before the High Court of Australia recognised the native title of the Meriam people of Torres Strait in the historic Mabo case. Taffe's (2018) sensitive analysis of the juxtaposition between Mary's telling of her father's heroic struggles in establishing his pastoral empire against tough odds, and her latter realisation of how this same colonialism brought about the suffering and destruction of Aboriginal traditional cultural life illustrates the internal and ongoing conflict confronting contemporary Australians as we recognise the damage of racial stereotyping, and discuss recognition of Australia's First People within our Constitution.

Wright (1991, p. 30) described Australia as a 'haunted country' cleft by 'the love of the land we have invaded and the guilt of the invasion'. 'The deepest crises experienced by any society are those moments of change when the current story becomes inadequate for meeting the survival demands of a present situation' (Berry 2015, p. xi), and, as will be discussed, the current story for the Traditional Owners of Northern Australia is woefully inadequate. It is essential we resolve this legacy of Australia's colonial past as a precursor to achieving truly sustainable development of our North, for as Massy (2017, p. 499) points out in his exploration of transformative change in Australia's extensive pastoral industries:

> Until there is true reconciliation and appropriate reparation and acknowledgement of the past injustices with Indigenous people, we modern non-Indigenous Australians won't be able to belong here; to truly set deep roots and our psychic selves in this ancient country; to truly become structurally coupled with the Australian earth.

Reconciliation needs to occur at national and individual levels. The former can be facilitated through political and community education and debate, but through this book I propose that, along with improved policy for North Australia's development, individual reconciliation will also benefit from a more complete understanding of an individual's operational context and their mechanisms for dealing with adversity.

1.1 Northern Australia's Existing Agricultural Capacity

Northern Australia is vast, over 40% of the mainland continent, geographically diverse, alternatively and, at times simultaneously, very wet and very dry (Lawn 2011). Precisely what comprises Northern Australia varies according to the context of the discussion, but here the description is aligned with that used by the 2015 *Our North, Our Future: White Paper on Developing Northern Australia* (Australian Government, p. 132): 'those parts of Western Australia and Queensland above the tropic of Capricorn as well as all of the Northern Territory'. It is a unique part of the tropics: with an enormous natural resource base yet remarkably small human population,[2] operating in a democratically governed first world economy.

This situation is at odds with international trends – almost half the world's population live in the tropics and tropical economies are growing 20 percent faster than the rest of the world (JCU 2014, p. xii). The world's population is projected to exceed 9 billion people by 2050, and this will be accompanied by an increase in the middle-class demographic (DAFF 2013) – Kharas (2010) predicts that more than half the world's middle class will be Asian by 2020. This larger, wealthier population will require more food and food of higher value. Linehan et al. (2012) speculate that the real value of global food demand in 2050 is expected to be 77 percent higher than 2007 levels, with most demand coming from Asia, particularly for meat and

[2] Around 1.3 million people – or 5% of Australians – live in the north (ABS 2015).

processed foods rather than traditional staple grains, and it is unlikely Asia will achieve food self-sufficiency (Andrews and Gunning-Trant 2013; Cole and Ball 2010). These projections have been widely interpreted by government and industry to provide opportunities for the growth of Australian agriculture, particularly in the underdeveloped north.

A viable and diverse agricultural industry operates in Northern Australia, with beef, sugar, dairy, corn, sorghum, peanuts, avocadoes, mangoes, nuts, chia, and a myriad of fruits and vegetables, as well as plantation timber including sandalwood. Agriculture is a major contributor to Northern Australia's economy, with a gross forecast value in 2015–2016 of $57.13 billion providing $43.4 billion in export earnings (ABARES 2015). For Far North Queensland, direct primary industry turnover in 2008–2009 was estimated at over $1.7 billion, with direct employment of about 9000 people (RDA 2011). Australia is the world's largest exporter of sheep and cattle, and 80% of exported cattle are from the north of the continent; and in 2015 AACo opened the Livingstone Beef facility – an abattoir near Darwin with eventual processing capacity forecast of 1000 head per day (McGauchie 2015).

Paralleling this diversity of agricultural production is the nature and scale of enterprises involved in production – from traditional family farm operations through to large corporate-owned operations, and from specialist peri-urban niche producers through to extensive leasehold grazing operations over hundreds of thousands of hectares. Further expansion of agriculture in Northern Australia is being actively promoted and arousing national interest, as demonstrated by release of the Australian Government White Paper on Developing Northern Australia in June 2015 and the 2013 Queensland Government commitment to a doubling of agricultural production by 2040 (Queensland Government 2013).

1.2 Why Consider an Individual's Resilience?

This book relates to the mechanisms and strategies primary producers adopt to manage the setbacks and adversity they face as part of farming in Northern Australia, particularly in the contemporary context enthusiastic for an expansion of Northern Australia's agricultural capacity. It does not intend to address the textural questions of which crops, what markets, or specific agronomic and animal husbandry practices related to any expansion of agricultural production. Rather, the focus is on the resilience[3] of individual producers involved in primary production in Northern Australia: the manner in which they cope with adversity and disaster,[4] and how it is

[3] An often used but loosely defined term, increasingly associated with community and individual response to adversity, and which will be considered at length in this book.

[4] The United Nations International Strategy for Disaster Reduction defines disaster as a 'serious disruption of a community or a societies functioning, causing widespread human, material, economic and/or environmental losses which exceed the ability of the affected community or society to cope using its own resources' (ISDR 2006, p. 5).

that under such conditions many actually prosper. Also included are perspectives on the influence of where we live and how we build, and important changes in the provision of health services across the northern landscapes.

This is social research and as such stands at the intersection of a number of disciplines in an attempt to ground individuals, and the events that affect them, into the broader social context of humanity. For, as Habermas (1972, pp. 155–156) points out:

> Life histories constitute themselves not only in the vertical dimension as a temporal connection of the cumulative experiences of an individual. They are also formed at every moment horizontally at the level of the inter-subjectivity of communication common to different subjects.

This focus is not intended to in any way diminish or devalue the importance to primary producers of access to technical information and advice or the need for ongoing research, development and extension efforts in agriculture, along with the provision and improved access to other services. Rather, the intention is to demonstrate that the use of such tools and information is not context-free and that an improved understanding of the contextual situation and world view of primary producers will have a demonstrable benefit in improved policy development and alignment, along with improving the individual producer's ability to consider and utilise such information.

My interest and enquiry into individual response started a long time ago and has strengthened with time. During my early professional career, I spent a decade undertaking solo biological surveys and then pest and weed management research and extension, in central and north Australia, repeatedly visiting many remote properties. This interaction provided the opportunity to get to know people, and the distances travelled provided time to reflect on their particular situation. I wondered what their life would be like to live – what would be the good aspects and what would be hard – and how might I cope if the roles were reversed. The thinking and perspective that flowed from such contemplations continued to occupy my thoughts long after the work was finished. I wondered what made farmers *tick*? It also fed my aspiration to *go farming* (though fiscal dictates required a start a few points removed from the extensive pastoralist industries in which I predominately worked).

While working as a pest management Extension Officer in the early 1990s, I was confronted by an incident illustrative of the occasional but very damaging disassociation between science practitioners and farmer's knowledge, specifically a scientist's contempt for the practical experience and opinion of farmers. While I knew this view was not universal, it made me consider how better outcomes could be achieved if these knowledge systems could work better together. This thinking resulted in my Master's thesis, which considered community participation in natural resource management (Noble 1997).

In 1997 I found an opportunity for affordable entry into farming: the tropical fruit industry, which was transitioning from a loose collection of individual growers into a cohesive industry with a promising future. The industry offered good returns from small acreage, it was situated in a beautiful part of the world (Queensland's

Wet Tropics), and while most people in Australia didn't know what a rambutan[5] was, everyone who tried one liked them. Industry members had a shared sense of purpose and values which aligned with my prior experience and expectations, and it was exciting! By late 2000, after a 3-year stint in Western Australia's Gibson Desert as inaugural Land Use Planner for the Ngaanyatjarra Council,[6] we (I was now married) had the wherewithal to buy the Tully property.

We embraced farming enthusiastically, confident that our professional skills could complement existing industry knowledge, and were instrumental in establishing a central packing shed to deliver economies of scale for smaller growers. Next came a grower-owned marketing company exporting fruit direct to Japan while providing coordination (and price stability) to the Australian market. Things were going well – the industry was growing, farm prices were rising as new people joined, and we felt our contribution was valued.

This sense of purpose disappeared early morning on 3 February 2011 when Cyclone Yasi cut a swathe through Far North Queensland's natural and production environments, compounding the impact from Cyclone Larry 5 years previous. Unlike Cyclone Larry, Yasi occurred in the aftermath of the Global Financial Crisis and in the context of a high Australian dollar and a 'summer of disasters' throughout Eastern Australia. The tropical fruit industry ceased to exist overnight apart from a few isolated remnants, along with coordinated marketing and farm values. The region's principal horticultural commodity, bananas, had 90% of production wiped out but 9 months later was facing chronic oversupply and below cost-of-production returns to growers as the cyclone *synchronised* return to production – a situation that persisted for many years.

In this same year, a ban on live cattle export to Indonesia,[7] compounded by widespread drought conditions, devastated Northern Australia's principal agricultural industry, extensive grazing, and I realised that whether an act of God or an act of parliament, a disaster was a disaster, and agriculture was particularly vulnerable.

It was while endeavouring to 'come to grips' with my personal post-cyclone situation that public discussion around developing Northern Australia and an expansion of Northern Australian agriculture came to the fore. I wondered how any industry expansion could be contemplated when the incidence of climate-related disasters was predicted to increase as a consequence of climate change (King 2010), with less climatic predictability and more disasters impacting upon an increasing and more

---

[5] Rambutan (*Nephelium lappaceum*) is the fruit of a medium-sized tropical tree in the family Sapindaceae, native to the Malay-Indonesian regions of tropical Southeast Asia.

[6] A Western Australian Aboriginal Corporation.

[7] In June 2011, the Federal Government suspended live cattle exports from Australia to Indonesia following the ABC's Four Corners programme on 30 May 2011 showing brutal slaughtering methods and mistreatment of cattle inside an Indonesian abattoir. While the ban lasted less than a month, its impacts continue to affect the beef industry today through a combination of narrowed market options compounded by drought conditions across much of Northern Australia. The unexpected market loss and resulting price plummet confused established drought management strategies of producers, and the impact of the ban extended to beef producers far beyond those directly dependent on the live export trade.

vulnerable population (ISDR 2008; Barratt et al. 2009; Handmer et al. 2012; Prabhakar et al. 2009), and particularly when regional and remote communities in tropical Queensland were thought to be among Australia's most vulnerable in the face of climate change (Dale et al. 2011). I wondered what made people stick to such a hard game and, more importantly, why some people appeared to not only survive adversity but to prosper.

This second decade of the twenty-first century is not the first-time grand plans, founded on the premise of underutilised land and bountiful water resources providing the wherewithal to feed the world's starving millions, have been advocated for Northern Australia. Consider though that contemporary farm returns in Australia often fail to meet actual production costs; the social, environmental and climatic unknowns associated with increased agricultural scale; and the inherent risks in agriculture, and suddenly the future does not seem quite so clear.

Agriculture is an industry connected across scale, commodity and community, and this innate capacity has delivered many demonstrations of innovation and success. Small growers benefit from larger growers' ability to attract infrastructure and service provision, but often smaller growers are the innovators who precede wider adoption. Industry collaboration has provided spectacular outcomes, such as recognition of the value of zebu cattle by Monty Atkinson in 1926 leading to development of the Droughtmaster beef breed; the 2004 industry-driven and world-first eradication of Black Sigatoka disease from North Queensland bananas; the whole-of-industry Brucellosis and Tuberculosis Eradication Campaign (BTEC) across Australia; and the ongoing community benefits provided by the Natural Disaster Relief and Recovery Arrangements response to twenty-first century cyclones and floods. With experience and persistence, past 'failures' are showing signs of success, for example, the sale of the Ord Stage II irrigation development to Chinese interests (Barnett and Grylls 2013), expanding banana production at Lakeland Downs (particularly post-cyclones), and dryland rice production in the Tully district of Queensland.

Can a growing Northern Australian agricultural sector better prepare itself for the natural, social, and economic pitfalls it <u>will</u> encounter along the way, embrace the concept of sustainable agriculture, and recognise and nurture the region's existing tropical agriculture expertise? Walker et al. (2010) propose featuring resilience and transformability alongside productivity as major objectives of research, as literature suggests that not only are resilient organisations about surviving but thriving; and the dynamic relationships between vulnerability, resilience, hazard impact, hazard change, adaptive capacity and social change in the context of climate change and disasters can inform approaches to planning for and developing community-based approaches to adaptation (Cottrell et al. 2011).

This importance of understanding an individual participant's perspective is supported by other researchers: for example, when considering reasons behind suboptimal adoption of seasonal climate forecasts by graziers for managing climate variability, Marshall et al. (2011, p. 514) found that social factors, not technical factors, were significant and hypothesise that:

> Factors that make resource-users dependent on natural resources (such as attachment to occupation and place, education, employability, environmental attitudes, local knowledge, and the quality and extent of formal and informal networks) act to influence resource-users in their decisions to adopt strategies that could enhance their capacity to cope and adapt to climate variability.

Improved understanding of how farmers think about and interact with their situation, along with clearer understanding of the interrelationships around this thinking, will provide opportunities for better planning and improved resilience of farmers dependent on climate-sensitive resources; and my experience told me this knowledge should be sourced directly, rather than filtered through a pre-existing theoretical construct.

The specifics or merit of what, how or where to farm should be left to the people and enterprises concerned, as the complexity and interrelationship of factors affecting success vary for every situation. However, there are factors affecting the capacity of individuals to establish and maintain successful agri-business enterprises that are influenced by government policy settings, which might be improved at minimal cost to deliver significant improvements in individual capacity to adapt to operational and social environments. These are not exclusively agriculture policies, as farming families also require education, health care, somewhere to live and somewhere to socialise; and better integration of diverse policies could improve outcomes and reduce costs.

Through a study of the context, personal strategies, perspectives and operating environment of individuals within Northern Australian agriculture (now and in the past), my objective was to identify and understand the factors and strategies that contribute to or enhance an individual's chance of achieving what they perceive as successful outcomes. It was thought that crisis or disaster could provide focal points to determine whether these factors and strategies could improve planning and policy outcomes and reduce industry risk.

Through a review of academic and grey literature filtered through personal industry experience and current research, I wanted to identify whether today's aspirations to expand Northern Australian agriculture differ from past attempts, and specifically what role an individual's approach and attitude might play in their success, or, to put it in the language of the participants, what made them 'stick at it'?

The end purpose of this book is to identify key elements and context, the consideration of which might assist planners and policy-makers to better engage with and deliver results for Northern Australian agriculture, as policy that is silo-based and implemented at scales removed from, or worse, without engaging with and understanding the local context, can lead to maladaptive responses.

1.3 Book Structure and Chapter Logic

The three author's personal experience (in agriculture, natural resource management, architecture and human health) in regional Australia informs the analysis of regional communities, industry trends and history; then focus this mix through the

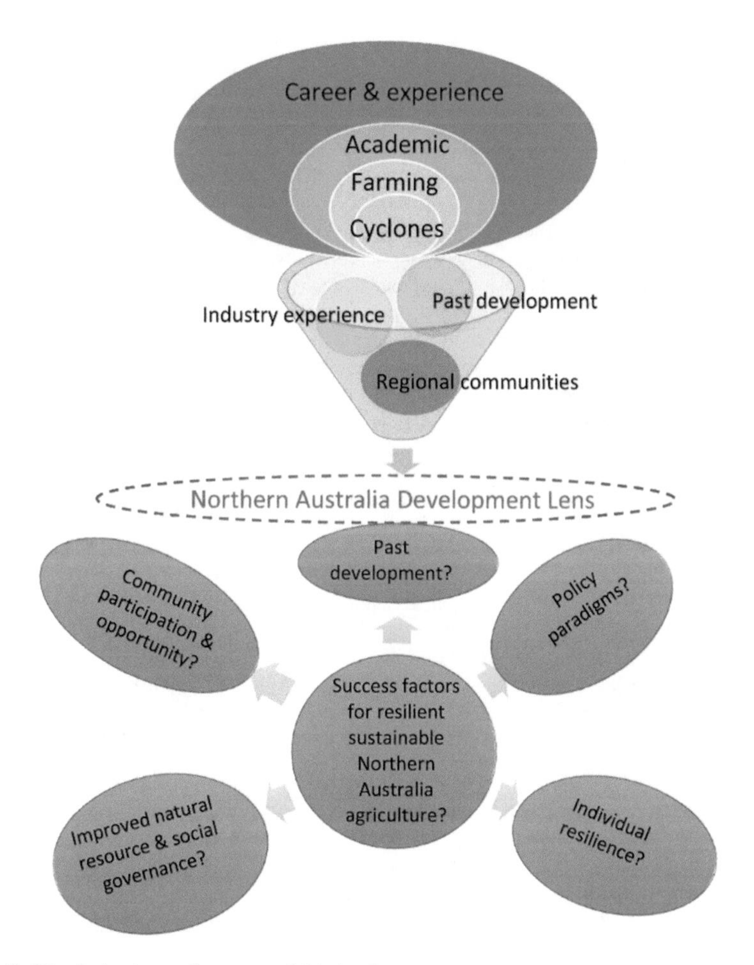

Fig. 1.2 The derivation and process of this book

lens of contemporary northern development aspirations with the intention of deriving a better understanding of what characteristics and processes enable individuals to survive and prosper in North Australian agriculture – their resilience (see Fig. 1.2).

There are three sections to this book. The first section, Chaps. 1, 2, 3, 4 and 5, gathers contextual information (historical and contemporary) important for understanding the resilience concept and how this relates to Northern Australia's place in the world. The second section, Chaps. 6, 7 and 8, is a consideration of why people, place and services matter, comprising analysis of individual resilience strategies, along with service delivery external to that specifically required for agriculture but essential to an individual's wellbeing – somewhere to live and personal health. In the third section of this book, Chap. 9 considers the interrelationships and synergies discussed previously and introduces a unifying thread. The concluding Chap. 10 reflects on the policy and planning context along with discussion of opportunities for incorporation of these concepts into planning and decision-making.

References

ABARES. (2015). *Agricultural commodities: September quarter 2015*. Retrieved from Canberra: www.agriculture.gov.au/abares/publications

ABS. (2015). *Regional population growth, Australia, 2013–14*. Retrieved from Canberra: www.abs.gov.au/ausstats/abs@.nsf/mf/3218.0

Andrews, N., & Gunning-Trant, C. (2013). *What Asia wants: Long-term food consumption trends in Asia*. Retrieved from www.agriculture.gov.au/SiteCollectionDocuments/abares/publications/longtermFoodConsumptionTrendsInAsia_v.1.1.0.pdf.

Australian Government. (2015). *Our North, our future: White paper on developing Northern Australia*. Canberra: Commonwealth of Australia. Retrieved from http://northernaustralia.infrastructure.gov.au/white-paper/files/northern_australia_white_paper.pdf.

Barnett, C., & Grylls, B. (Producer). (2013). Ord development takes important step. Retrieved from www.mediastatements.wa.gov.au/pages/StatementDetails.aspx?listName=StatementsBarnett&StatId=7425

Barratt, P., Pearman, G., & Waller, M. (2009). Climate change: What sort of resilience will be required. In S. Cork (Ed.), *Brighter prospects: Enhancing the resilience of Australia. Australia 21, shaping the future* (pp. 30–35): ANU and University of Melbourne.

Bennett, M. M. (1927). *Christison of Lammermoor*. London: Alston Rivers Ltd.

Berry, T. (2015). *The dream of the earth*. Berkeley: Counterpoint Press.

Cole, M., & Ball, G. (2010). Global trends and opportunities in food and nutritional sciences. *Food Australia, 62*(10), 460–464.

Cottrell, A., King, D., & Dale. A. (2011). *Planning for uncertainty: Disasters, social resilience and climate change*. Paper presented at the 3rd World Planning Schools Congress, Perth, WA. http://eprints.jcu.edu.au/18771/

DAFF. (2013). *National food plan, our food future*. Canberra Retrieved from daff.gov.au/nationalfoodplan

Dale, A., Vella, K., Cottrell, A., Pert, P., Stephenson, B., King, D., …, Gooch, M. (2011). *Conceptualising, evaluating and reporting social resilience in vulnerable regional and remote communities facing climate change in tropical Queensland. Marine and Tropical Sciences Research Facility (MTSRF) Transition Project Final Report*. Retrieved from Cairns: http://www.rrrc.org.au/publications/research_reports.html

Gammage, B. (2011). *The biggest estate on earth: How aborigines made Australia*. Crows Nest: Allen & Unwin.

Habermas, J. (1972). *Knowledge and human interests* (J. J. Shapiro, Trans.). Boston: Beacon Press.

Handmer, J., Honda, Y., Kundzewicz, Z. W., Arnell, N., Benito, G., Hatfield, J., et al. (2012). Changes in impacts of climate extremes: Human systems and ecosystems. In C. B. Field, V. Barros, T. F. Stocker, D. Qin, D. J. Dokken, K. L. Ebi, M. D. Mastrandrea, K. J. Mach, G.-K. Plattner, S. K. Allen, M. Tignor, & P. M. Midgley (Eds.), *Managing the risks of extreme events and disasters to advance climate change adaptation. A special report of working groups I and II of the intergovernmental panel on climate change* (pp. 231–290). Cambridge: Cambridge University Press.

Huxley, T. H. (1863). *Evidence as to man's place in nature/by Thomas Henry Huxley*. London: Williams & Norgate.

ISDR. (2006). *Disaster risk reduction: 20 examples of good practice from Central Asia* (DRR10628). Retrieved from http://www.unisdr.org/files/2300_20GoodExamplesofGoodPractice.pdf

ISDR. (2008). *Climate change and disaster risk reduction*. Retrieved from Geneva, Switzerland: http://www.wmo.int/pages/prog/dra/vcp/documents/7607_Climate-Change-DRR.pdf

JCU. (2014). *State of the tropics: 2014 report*. Retrieved from Townsville: www.stateofthetropics.org

Kharas, H. (2010). *The emerging middle class in developing countries* (Working Paper No. 285). © OECD Development Centre.

King, D. (2010). Issues in identifying climate change adaptation within community hazard mitigation. *International Journal of Emergency Management, 7*(3–4), 269–280. https://doi.org/10.1504/IJEM.2010.037011.

Lawn, B. (2011). *Opportunities and conflict in agriculture and NRM in the Australian tropics*. Paper presented at the "A food secure world: Challenging choices for our North" The Crawford Fund 2011 State Parliamentary Conference, Parliament House, Brisbane. www.crawfordfund.org/assets/files/conference/Queensland/Bob_Lawn_ppt.pdf

Linehan, V., Thorpe, S., Andrews, N., Kim, Y., & Beaini, F. (2012). *Food demand to 2050: Opportunities for Australian agriculture*. Paper presented at the ABARES conference, Canberra.

Marshall, N. A., Gordon, I. J., & Ash, A. J. (2011). The reluctance of resource-users to adopt seasonal climate forecasts to enhance resilience to climate variability on the rangelands. *Climatic Change, 107*(3–4), 511–529. https://doi.org/10.1007/s10584-010-9962-y.

Massy, C. (2017). *Call of the Reed Warbler: A new agriculture, a new earth*. St Lucia: University of Queensland Press.

McGauchie, D. (2015, February 21) *Historic day for northern cattle industry: AACo officially opens its livingstone beef facility near Darwin/Interviewer: M. Brann*. ABC Rural Radio.

Noble, K. (1997). *The oft-forgotten human dimension of feral animal research and management*. Townsville: (MSc), James Cook University.

Pascoe, B. (2014). *Dark Emu. Black seeds: Agriculture or accident?* Broome: Magabala Books.

Prabhakar, S. V. R. K., Srinivasan, A., & Shaw, R. (2009). Climate change and local level disaster risk reduction planning: Need, opportunities and challenges. *Mitigation and Adaptation Strategies for Global Change, 14*, 7–33.

Queensland Government. (2013). *Queensland's agriculture strategy: A 2040 vision to double agricultural production.*

RDA. (2011). *Regional road map – Far North Queensland & Torres Strait Region 2011–2012*. Retrieved from http://www.rdafnqts.org.au/images/pdf/RDA%20FNQTS%20Regional%20Road%20Map%202012%20-%20FINAL%20comp.pdf

Taffe, S. (2018). *A white hot flame: Mary Montgomerie Bennett – Author, educator, activist for indigenous justice*. Clayton: Monash University Publishing.

Walker, B., Sayer, J., Andrew, N. L., & Campbell, B. (2010). Should enhanced resilience be an objective of natural resource management research for developing countries? *Crop Science, 50*(Supplement 1), S-10–S-19. https://doi.org/10.2135/cropsci2009.10.0565.

Wright, J. (1991). *Born of the conquerors*. Canberra: Aboriginal Studies Press.

Chapter 2
Resilience Discourse and Adaption Strategies

Humanity looks to the future, and wants some of what it now values to be there
Juech and Martin-Breen (2011)

2.1 Introduction

This chapter explores the extent and variation of resilience concepts and theory across many fields of science, which is then related to popular usage of the term to explain why an appreciation and reconciliation of the two ontologies are important for improved governance and policy development within agriculture in an increasingly globalised world.

2.2 A Resilient Concept

Resilience often appears with words such as sustainability, vulnerability and robustness, though the relationships between these terms are loose and used to describe 'particular ends rather than theoretical constructs' (Martin-Breen and Anderies 2011, p. 47). Reghezza-Zitt et al. (2012, p. 1) describe resilience as a 'fashionable concept' which is 'now a must in both academic research and management', a view which could be supported by Xu and Marinova's 2013 analysis which is identified from the Web of Science database more than 900 cited papers published on the topic of resilience since 1973, with a strong upward trend.

Notwithstanding the influence of fashion, Martin-Breen and Anderies suggest there is 'important information captured by studying resilience, information that is traditionally not studied or left out' (p. 52), and that in studying individual objects including persons 'using resilience can fill a knowledge gap' (p. 50). This view is supportive of the premise by Ungar et al. (2007) that tensions are dynamic, converg-

Keith Noble has contributed more to this chapter.

K. Noble et al., *Agriculture and Resilience in Australia's North*,
https://doi.org/10.1007/978-981-13-8355-7_2

ing in different ways across time and throughout a person's life, and that it is at the intersections of these tensions that what constitutes resilience in any given culture or context is revealed: resilience is about finding a way to live in relative comfort despite contradictions and conflicts, to continue to navigate and negotiate challenges. Unger et al. believe that in this way, resilience is not a permanent state of being, rather it is a condition of becoming better. While conventional adaptation approaches emphasise adjustments to move towards a desired state that reduces risks, resilience puts greater attention on building capacity to cope with future change (Brown 2016).

These views are fundamental to my work: In many instances, agencies are locked into short-term planning and project cycles and do not have the capacity and resources or resolve to look at longer-term changes and drivers. I wanted to develop an improved understanding of what constitutes resilience in Northern Australian farmers, which in turn could assist policy alignment supportive of individuals and their communities for the benefit of all Australians.

2.3 Conceptualising Resilience: Origins and Evolution

Resilience theory in ecological systems is a well-developed and quantifiable field of science describing the capacity of a system to absorb disturbance and reorganise while undergoing change so as to still retain essentially the same function, structure, identity and feedbacks (Walker et al. 2004). Social resilience on the other hand is described by Barrett (2013) as presently being more 'an ubiquitous buzzword with a lot of arm-waving', but he does see 'big opportunity for ecologist – economist collaboration' in this area.

In an exploration of the existing literature on resilience with respect to disasters, Boon et al. (2012, p. 384) found the earliest studies to use the term resilience appear to be the 1940s work of Garmezy, Werner and Smith which focused on understanding the development of psychopathology in *at-risk* children, predating the widespread adoption of the term into ecosystem science. The closely related term of robustness (also used in psychology) is more often applied to designed rather than predominately self-organising systems, where resilience is more commonly used (Martin-Breen and Anderies 2011, p. 47).

Zhou et al. (2010, p. 22) identified 28 definitions of social resilience, with many of the discrepancies in meaning 'arising from different epistemological orientations and subsequent methodological practices'. Skerratt (2013, p. 36) explains that although international literature does not share a consensus in the definition of resilience or of community resilience, the spectrum of literature on rural community resilience is 'wider than bounce-back from external shock', and Kaplan (2009, p. 64) observes that when discussing resilience in this context 'competence in resolving issues in one developmental period does not predict later competence in a linear, deterministic way'. Ann S. Masten and Obradovic (2006, p. 22) caution investigators that 'resilience definitions are embedded in cultural, developmental, and historical contexts, even if these contexts are assumed rather than made explicit'.

There are however terms that have acquired specific meaning within the social resilience literature around the concepts of adaptation and coping. *Adaptation* is accepted as meaning adjustment in the face of change and can be positive, negative or neutral and is a key concept of evolutionary theory, though it differs from *adaptive capacity*. Adaptive capacity can be increased through adaptations that accurately and effectively respond to *events*. This is different from a *coping strategy*, which can be seen as an iterative process of individual, intentional change in response to a stressor, sometimes viewed as a low-level adaptive behaviour (Dale et al. 2011).

Martin-Breen and Anderies (2011) were commissioned by the Rockefeller Institute to undertake a rigorous literature review on resilience and concluded that within the scientific domain that focuses on the interactions between people and environment, resilience has evolved into an intellectual framework for understanding how complex systems self-organise and change over time, though the relationships are still quite loose and often used to describe particular ends rather than theoretical constructs. They found that in complex adaptive systems, resilience is best defined as the ability to withstand, recover from and reorganise in response to crises: 'Function is maintained, but system structure may not be' (p. 7).

Resilience can also be generally defined in two broad ways, as a desired outcome(s) and as a process leading to a desired outcome(s) (Kaplan 1999; Winkworth et al. 2009), and as such can be investigated at any level: individual, community, organisation or ecosystem, though 'no single level is the "correct" one for analysis' (Zhou et al. 2010, p. 30). In fact, the level one chooses for investigation depends on 'the issue or question of interest. Conceptually, the simplest level of investigation is individual resilience' (Boon et al. 2012, p. 385). Zhou et al. (2010, p. 26) ask:

> Is resilience the opposite of vulnerability? Is resilience a factor of vulnerability? Or is it the other way around? It is not easy to provide single answers to these questions. Addressing this relationship is important in defining the meaning, implications and applications of resilience.

With respect to individual resilience, most studies indicate it to be partly a trait and partly a dynamic process and that it can be promoted by two groups of generic factors:

1. Personal attributes such as social competence, problem-solving, autonomy, self-efficacy and sense of future and purpose
2. Contextual environmental influences such as peers, family, work, school and local community

Through an article-screening process of disaster resilience literature, Boon et al. (2012) also found that individual resilience is described differently from community resilience and that there is a challenge in relating the two because 'existing models are from either the psychological or sociological perspective, but without an integration of the two' (p. 389). Furthermore, while resilience within an individual is believed to be a process rather than a steady state, 'with a person's level of resilience potentially varying over their lifetime', many authors 'emphasise the importance of recognising the dynamic, interactive nature of resilience and the interplay between an individual and their broader environment' (p.385–86). However, after their extensive consideration of the literature, Boon et al. concluded that, 'Research linking individual to community resilience is very scarce worldwide and non-existent in Australia', which is problematic because 'developmental science and ecological science perspectives intersect to explain resilience at both individual and community levels' (p. 402–403).

Martin-Breen and Anderies (2011) state that in terms of research 'most resilience research outside of the psychological tradition stops at the community level' (p. 36), which is again problematic because what households (individuals) do will map up to higher levels in unpredictable ways, particularly since local vulnerabilities are nested and teleconnected by globalisation, so 'resilience thinking is valuable in framing and discussing aspects of sustainability and sustainable development' (p. 13).

Certainly psychological growth can occur as the result of living through adversity, and this can assist the individual when future stressors are encountered (Polk 1997), and it is this assumption of post-stress growth that leads Aldwin (2007) to deduce resilience to be more than stoicism or survival. But it is important to remember that 'resilience is not a steady state in an individual' (Hegney et al. 2007, p. 11) – it can vary throughout a person's life, being more of a process than a steady state (Winkworth et al. 2009), or even a 'complex family of concepts' beyond a single trait or process, and to which there can be 'multiple pathways' but no 'magic bullets' (Ann S. Masten and Obradovic 2006, pp. 22–23). A. S. Masten and Obradovic (2008) also state that this complex interaction of personal resilience with the environment and particularly the variation over lifespan is rarely researched.

Carl Folke et al. (2002, p. 437) describe accumulating evidence from diverse regions around the world that both 'natural and social systems behave in nonlinear ways, exhibit marked thresholds in their dynamics, and that social-ecological systems act as strongly coupled, complex and evolving integrated systems'. They suggest that a multilevel governance system is required that will allow for learning and increase adaptive capacity without foreclosing future development options. This distinction is well explained by Nelson et al. (2007): while *adaptation* is a process or activity undertaken in order to alleviate adverse impacts or take advantage of new opportunities, *adaptive capacity* consists of the preconditions necessary to enable adaptation to take place – that is, it is a *latent characteristic* which must be activated in order to affect adaptation.

Humans can reason and choose (Walker and Salt 2006), and a resilient individual might stop what they are doing and do something completely different to survive. They might transform rather than simply cope into a new crop, conservation management, tourism or off-farm work. They might stop farming altogether. This is not the same as crossing an ecological threshold which the system then settles permanently into. This is supported by A. S. Masten and Obradovic (2008, p. 13) who recommend that:

> knowledge from research on human resilience from the developmental and behavioural sciences must be integrated with knowledge about resilience from research on many other components of the complex interacting systems in which human life is embedded.

So while Cottrell et al. (2011) warn that a shared view of what constitutes resilience in general and community/social resilience in particular is likely to remain elusive, she points out that it is essential approaches taken to planning are context-specific and developed in conjunction with those people who are most affected. Flint and Luloff (2005) point out that, 'disaster research tends to focus on the immediate post disaster experience … and does not routinely study the long-term recovery path' (p. 402). They conclude that to improve policy- and decision-making, it will be important to properly uncover the perceptions, capacities and range of contextual variation within community response to disaster and that an important component of achieving this will be adaptive qualitative and ethnographic research, as the 'ways risks are perceived within communities influence the range of actions undertaken to reduce them' (p. 408). Martin-Breen and Anderies (2011, p. 35) also support the application of resilience to governance and management and make recommendations which 'highlight how resilience can stimulate both new ways of understanding "wicked problems[1]" and new ways of developing solutions to them'.

Finally, Reghezza-Zitt et al. (2012) suggest that we must also study resilience as a political line since the term has now entered the political lexicon: showing successful reconstruction as quickly as possible is a strong political move that shines a good light on leaders. In such a situation, 'Resilience's key issue is to know who says that there is resilience, when and why' (para. 61), as in society there are always rewards and punishments, and 'resilience for some people or places may lead to the loss of resilience for others' (Davoudi et al. 2012, p. 306). A bonus of this could be that such use of the term in these situations 'highlights the necessity of getting over "zero risk" logic' (Reghezza-Zitt et al. 2012, p. 66).

Vignette 2.1: The Cutten Brothers and Climate Change
April Fool's Day, 1882:

Four brothers row a borrowed boat down the Tully River (Far North Queensland) and north along the coast in search of land. James, Leonard, Sidney and Herbert Cutten had arrived in Sydney a decade previous and gathered considerable experience in their journey north. Though tired when they beached their 'flattie', the long hard shining sands of Mission Beach with their inviting shade of cool, fragrant calophyllums carried James's surveyor eye on to the rich vistas of gently rolling well-watered land; with forests of cedar, nutmeg, quandong and native ginger framed by the majestic peak of Clump Mountain and the sheltered Shangri-La of Bingil Bay. It was a strange exotic land to these young Englishmen, but they named it Bicton in memory of pleasant times in the English Bicton Hills. Coincidently, it was also the local aboriginal word for 'good camping ground and plenty of fresh water'.

---

[1]A wicked problem is a problem that is difficult or impossible to solve because of incomplete, contradictory and changing requirements that are often difficult to recognise. The use of the term *wicked* has come to denote resistance to resolution, rather than evil (Wikipedia contributors 2015).

What followed was hard hot work in the steaming jungle with axe and saw, but the four were a powerful team. Pineapples and bananas were planted with assistance from the incumbent Aboriginal population of circ. 400 people (a homogenous tribe quite separate from the nearby Mission Beach people). Tea, coffee, chicory, coconuts, citrus, mangoes, tobacco, spice and jack-fruit followed; along with a stone breakwater which enabled the now regular coastal shipping to safely load produce for southern markets. By 1891 the coffee harvest justified purchase of a mill and expansion to 100 acres and, assisted by seasonal Aboriginal labour, produced a quarter of a million pounds of ground coffee annually for export to London.

They prospered, the house grew: there was a two-story packing shed, sawmill, case-mill, a wooden railway to the stone wharf, and marriage for James. E. J. (Ted) Banfield, the British journalist turned Dunk Island resident and author of "*Confessions of a Beachcomber*" (London, 1908) was a regular visitor, and his writings evoke the idyll of the tropical life they shared. The brother's fortunes stood at their zenith when their mother died in 1908, and Banfield read the funeral service.

But trouble started when Chinese banana growers started paying Aboriginal labour with opium, resulting in the government limiting employment to three '*boys*' only when the Cuttens needed 70 for their operation. The Yongala cyclone of 1911 damaged crops, orchards, breakwater and boats; and without their earlier Aboriginal labour the Cutten brothers (now in their 60s) mortgaged the property for repairs. Another cyclone in 1913 wreaked further damage. No sooner were repairs effected than WWI broke out and Bicton's lifeline was cut as all ships were diverted to war service. It was impossible to get produce away, so the brothers pulled in their belts and tried cutting timber. This proved a failure, but a greater disaster was Sidney being badly injured whilst scrub-felling to satisfy their property terms under the Land Selection Act. Then, in 1918, the great cyclone ended the Cutten's 40-year époque at Bingil Bay.

The Cuttens had survived cyclones in 1890, 1911 and 1913, but the cyclonic surge of 10 March 1918 carried 2 miles inland, smashing the stone breakwater and boats in an instant. The 13-room house disintegrated in the 200 miles per hour winds along with sheds, crops, orchards, and 200,000 super feet of sawn timber ready to be shipped. Of the Cutten empire nothing remained, though miraculously everyone at Bicton survived; unlike the township of Innisfail and surrounding areas where almost 100 died. Banfield recorded it as the cyclone of the century and the greatest natural disaster to hit Australia's east coast.

[Extracted from R. J. Taylor *The Lost Plantation: A History of the Australian Tea History* (Cairns: G. K. Bolton 1982)]

The 1918 tropical cyclone destroyed the Cuttens enterprise, but it did so on the back of three earlier cyclones, labour and freight shortages, a world war, legislative changes and advancing age. Whether this cyclonic event, or Cyclones Larry and

Yasi 100 years later[2] were attributable to climate change, was of secondary consequence to those who suffered their impact.

2.4 What's in a Name: Theory vs the Vernacular

Why look at resilience? Why is the concept important? As described, resilience has been used in many fields of science over the past seven decades, and it is now also found in political science, business administration, sociology, history, disaster planning, urban planning and international development. The shared use of the term does not, however, imply unified concepts of resilience, nor the theories in which it is embedded. Different uses generate different methods, sometimes different methodologies. Evidential or other empirical support can differ between domains of application even when the concepts are broadly shared (Martin-Breen and Anderies 2011). The use of resilience across such varied fields does however indicate its appeal; and perhaps it has become a 'metaphor' for something that cannot be directly observed or measured.

While this might be an interesting and important deliberation academically, from a farmer's perspective, a disaster is a disaster, irrespective of whether it was a consequence of an act of God or an Act of Parliament; and the eventual impact of any event on an individual and their enterprise will be determined by myriad interconnected factors. While a return to *normal* or *bounce back* is probably desirable after a natural hazard like a cyclone, 'this view focuses on the short-term restoration of functions and essential infrastructure' (Darnhofer 2014, p. 466), whereas a more desirable outcome from a medium to long-term view might be for a *bounce forward* or transformative outcome, which Darnhofer argues is a property which plays a 'more important role in social systems than in ecological systems, especially when taking a long-term view' (p. 466).

There is also a third, more radical transformative dimension to resilience, where change occurs to take advantage of new possibilities and opportunities and 'positively develop and to thrive in the face of change and uncertainty' (Brown 2016, p. 11). So if resilience is the ability of an individual or social system to persist, its study will need to include both adaptive and transformative capacities – to *bounce back* and *bounce forward* along with the 'three heuristics of resilience management … resilience, adaptability and transformability' (Dale et al. 2011, p. 34).

Would such resilience be an inherent attribute of a system, or should it be considered a process? While these two positions are not necessarily opposed, they have different methodological and theoretical implications when attempting to translate theory into operational terms, particularly when the term is used by non-scientist decision-makers (Reghezza-Zitt et al. 2012). Flint and Luloff (2005, p. 408) believe that 'when disaster strikes, regardless of origin, local capacities for action and the

[2] The cumulative effects of 2006 Cyclone Larry and 2011 Cyclone Yasi all but destroyed Queensland's emerging tropical fruit industry.

different ways they manifest need to be incorporated in our studies'. However, Flint and Luloff point out that in the literature there is often a bifurcation between presumed origins of risk, with many researchers distinguishing between technological and natural hazards on the assumption that natural risks are met with *therapeutic* community characteristics that serve to ameliorate disruption, whereas technological risks are assumed or found to be met with *corrosive* community processes that break down the social fabric due to the ambiguity of risk. However, little empirical evidence exists to support such assumptions, and distinguishing between technological and natural risks could promote a false dichotomy. In a large geographic region like Northern Australia, and with the diversity that comprises its agricultural industries, it would be unlikely a natural disaster would uniformly impact the entire region or all producers of a specific commodity, whereas the anthropogenic 2011 ban on live cattle export to Indonesia did impact cattle producers across the entirety of Northern Australia, and the flow-on impacts affected the entire regional economy for a very long time.

So how do communities and individuals respond to risks and disasters, and how important is the nature of the impact? Obviously, not all groups or individuals within a community will be equally exposed to risks or disasters, and individual response capacity will be affected by the difficulties and conditions of everyday life, along with the legacy of resource interdependence in natural resource-based communities. However, it is the study of actors in their associational action at the community level of analysis that can move the study of risk and disaster towards an understanding of the variations in response and recovery (Flint and Luloff 2005).

We are paying attention to the resilience agenda now because it has become policy, but policy must be measured. The question that must be addressed even before that of *how do we measure it* is exactly *what is it we wish to measure*? When the foundation for modern medicine was being constructed as part of the seventeenth-century revolution of natural sciences, a mechanistic world view was dominant, and humans were understood as complex machines ruled by the natural laws that regulate everything else in nature. As mathematics was considered the cradle of science, that which could be examined numerically and described by statistics was deemed to have higher scientific value (Dahlberg 2013). In the same way, the application of ecological resilience theory to individuals and communities risks an extension of this view, which is a risk when a highly defined term in one context is applied to another. An associated risk is that by overstating the parallels between ecological and social systems, governance and power issues that affect social change can be overlooked, leading to Cote and Nightingale (2012) arguing that the focus of research should move from the *content* to the *context* of knowledge production.

Despite all the above, and recognising the academic complexity around theoretical underpinnings of the term, neither the media nor community are likely to stop using *resilience*, particularly when discussing farmers and adversity. It is a term that is popularly understood even if an exact definition remains elusive. As Lélé (1998, p. 249) observed, 'Resilience is turning out to be a resilient concept'.

Latour (2005) argues that to be objective, social science researchers should not muffle informants' precise vocabulary into an all-purpose metalanguage but, rather, describe what people say and do. In recognition of this, while an appreciation of the context being discussed is important, it is equally important that we not abandon the term or invent another and that the science be reconciled *as much as possible* with community understanding.

2.5 Resilience, Adaptive Capacity and Transformation in Agriculture

Humanity and its familial and communal structures have, like ecosystems, been around a very long time and have persisted despite adversity. They are, in a broad sense, naturally resilient (Martin-Breen and Anderies 2011). Natural resource-based industries like agriculture occupy a unique interface between society and the environment: farmers generally operate as part of extended supply chains within the vagaries of global markets, and as a consequence, both these industries and their communities often appear particularly vulnerable to the negative effects of environmental and social change, particularly environmental risks and disasters (see Gaventa 1980 in Flint and Luloff (2005, p. 399)).

Certainly, individual farmer's vulnerability exists and is real, alongside their resilience, and often the weighting represented depends on the focus of the writer. Over and above extreme circumstances or disaster events, it is evident that individuals living in rural communities face 'life circumstances and unique ecologies which differ markedly from populations living in urban centres' (Hegney et al. 2007, p. 3). It is not surprising then that much risk and disaster research considers farmers as victims of destructive environmental and societal processes who are:

> limited in their ability to better position themselves to address negative circumstances …. [and] trapped in endless cycles of vulnerability and in constant need of external assistance for development, risk mitigation, and disaster recovery. (Flint and Luloff 2005, pp. 399–400)

The work by Flint and Luloff (2005) on improved understanding of natural resource-based communities and their risks and disaster experiences is driven by a commitment to improving policy- and decision-making, and they point out that:

> Natural resource-based communities are dynamic places embedded within an environmental and geographic context and the homes where people with collective, intersecting, and competing values interact. Though the legacy of dependency remains an obstacle, viewing natural resource-based communities as only being vulnerable is insufficient and flawed. Community capacity exists despite vulnerabilities. How people work together to make decisions and act on their collective and intersecting values in the face of risks, disasters, and everyday vulnerabilities is central to investigations in these areas. *(p. 400)*

Consequently, and in the interests of better policy outcomes, Flint and Luloff encourage the study of community, risk and disaster 'beyond a simple and exclusive focus on vulnerability' to assess broader community sentiments and responses (p. 408). This is supported by C. King (2008) who, when investigating agri-ecological systems, proposed that it is the diversity of function of systems at multiple scales that enhances both ecological and community resilience.

Martin-Breen and Anderies (2011, p. 10) recommend that, when considering the broader implications of risk and the appropriate allocation of limited human resources to address it, one needs to answer not only the questions 'resilience of what?' and 'resilience to what?' but also the important question of 'resilience for whom?' They highlight the importance of adaptive capacity in increasing the resilience of a system, as adaptive capacity requires adaptation processes that are both anticipatory and effective 'in creating systems that are able to maintain their state in response to unexpected crises' (p. 48).

Agriculture is a risky business. The endless variables that farmers must evaluate and plan for include weather, markets, supply chain, finance, labour availability, changing legislation, natural disasters, pests and disease. This limits farmer's ability to plan, as 'much is unknown and unable to be known, where great uncertainty prevails, and much is uncontrollable' (Malcolm 2004, p. 399). Darnhofer (2014) builds on this reality to understand resilience as encompassing buffer, adaptive and transformative capability. She argues that resilience thinking offers alternative insights into farm management and how farmers can achieve balance between short-term efficiency and long-term transformability while balancing exploitation and exploration. She states that while farm resilience can be strengthened or eroded by policy measures and family dynamics 'overall resilience proposes an alternative conceptual lens to one building on equilibrium, thus highlighting complex dynamics and the role of farmer agency in navigating change' (p. 461), because 'we are regularly confronted with events even experts and dedicated institutions failed to anticipate, highlighting the difficulties of prediction and the limits of focusing on known risks' (p. 476) when operating in a Newtonian world view[3].

2.6 Resilience and Policy Paradoxes

The application of neo-liberal policy in Australia has deliberately sought to foster self-reliance in the management of environmental risk by Australian agriculture rather than expecting it to be addressed through government funding as a national problem (Lawrence et al. 2013). However, while farmers have accepted their responsibility to manage risk, their capacity to do so is often sorely tested (Gill 2011). Consider that as a consequence of constantly dealing with risk, farmers are often considered by other sectors of society as conservative when it comes to issues like politics, and it is 'understandable that farmers are cautious and contest the claims of those who would have them reorganise current production systems' (Lawrence et al. 2004, p. 256).

---

[3]A view which considers the world as an orderly mechanical device whose behaviour can be explained and predicted by mathematical rules and monitored by using a command and control approach.

Parallel to Lawrence's description of the increasing expectation by government of self-reliance in the management of environmental risk by Australian agriculture, Brennan (2008) describes how both America and the European Union have witnessed a consistent devolution of responsibility for social and other services from the government to the local level, with communities being called upon to do more with less. Brennan goes on to describe how community agency, or the capacity for local action and resilience, can emerge as a consequence and that this agency can be seen as the capacity of people to manage, utilise and enhance the resources available to them. This shift in focus occurred through recognition that 'local residents are better suited to address their immediate and long-term needs, as well as being intimately familiar with mechanisms for achieving these' (p. 60). Brennan states that both research and applied programs and policies that shape community adaptive capacities can emerge 'based on the premise that active communities have the capacity to improve local well-being and directly shape their resilient capacity' (p. 61).

Early Australian settlers lacked the skills and knowledge of their new environment to realise that their introduced European agricultural practices were often unsuited to Australia (Gray and Lawrence 2001). Governments used pastoral lease conditions to facilitate and encourage land use intensification and closer settlement, an approach often incompatible with the unreliable climate and limited carrying capacity of the rangelands (Productivity Commission 2002). The resulting negative impact of agriculture on Northern Australia's natural environment and biodiversity has been extensively documented, and climate change is predicted to bring new pressures to bear on both agriculture and biodiversity (Cocklin and Dibden 2009).

The International Assessment of Agricultural Knowledge, Science and Technology for Development report (2009a) describes agriculture as a multi-output activity producing not only commodities but environmental services, landscape amenities and cultural heritage. A major challenge facing humanity is the continued production of food and fibre for a burgeoning population while avoiding long-term negative impacts on ecosystems and the services they provide (Folke et al. 2010; Rist et al. 2014; Walker 2014). However, production ecosystems

> typically have a high dependence on supporting and regulating ecosystem services and while they have thus far managed to sustain production, this has often been at the cost of externalities imposed on other systems and locations (Rist et al. 2014, p. 1)

– a situation they describe as 'coerced resilience'. They therefore suggest expanding the resilience framework to accommodate the specific characteristics of production systems. Their focus is however on ecosystem (engineered) resilience (see Martin-Breen and Anderies), not the capacity of the farmer to achieve such transformation.

While McManus et al. (2012, p. 28) warn policy-makers that 'perceptions of the local economy, environment and community are inter-related and resilience is dependent on all three simultaneously', there are many examples where societal pressure to integrate social issues into natural resource management planning has resulted in a technocratic response (the identification of indicators) rather than a strategic approach to planning for improvements in the state, trend or resilience of the asset, an example being the extended policy conflict and delays experienced in the Murray-Darling Basin water management planning, though anecdotal evidence would suggest 'the problem is as much due to the lack of exposure and influence of the social science community within the natural resource domain versus the domination of the biophysical sciences in that domain' (Dale et al. 2011, p. 56).

Cocklin and Dibden (2009) postulate that it is possible to envisage mitigation and adaptation responses that would alleviate pressures on all three systems (climate, agriculture, biodiversity), and there has been a wave of interest in shifting emphasis away from productivity enhancement and towards sustainability and resilience (IAASTD (2008, 2009b); McNeely and Scheer (2003). This movement is being assisted and facilitated in Australia by Regional NRM bodies[4], who are improving participation and shared understanding by all regional land managers in natural resource management (personal observation and experience over the past decade).

Farmers are popularly perceived as resilient (Hodgkinson (2014); Landry (2014); Neales (2015)) and often self-identify as being resilient (Cristaudo (2012); McFarland (2013)) – capable of handling the uncertainty and adversity that life on the land brings. But farmers are also members of regional communities and broader society, from whom they derive and contribute support. Farmers interact with and transition in and out of all aspects of society, and this can introduce risks that are beyond the control of individual farmers, no matter how prepared. While it might not be a policy or societal intention, increasing socio-economic marginalisation could be an inadvertent or perverse outcome of inadequately considered policy such as the 2011 live cattle export ban, with farmers at a loss to understand the thinking and actions of the animal rights lobby and the subsequent government decision, and ill-prepared to address the ramifications.

2.7 Future Trends and Direction

Worldwide, there is an emerging discussion around the need to explore alternate governance systems from global through to local scales in order to both address the social, economic and environmental challenges facing today's world and to improve multisector cooperation, particularly since *command-and-control* regulation has been found wanting (see Taylor 2010).

Higgins et al. (2010) describe contemporary society's global shift from public to private forms of governance and how farmer-initiated environmental management systems (EMS) often take a proactive approach to environmental issues in order to avert more onerous legislative intervention by government. As a corollary of this, Allan Dale et al. (2013) describe how linear governance systems that are poorly integrated with the wider system can constrain thinking, have limited benefit and

[4]There are 56 regional NRM organisations across Australia that act as delivery agents under the regional stream of the National Landcare Program – see www.nrm.gov.au/regional/regional-nrm-organisations.

even be counterproductive. They describe a systemic/adaptive means to optimise collaborative effort and have named it *Governance Systems Analysis*.

Sayer (2010, p. 20) extends this argument in the developing world by cautioning environmentalists about resisting agricultural innovations that may have short-term or local negative impacts on nature but which might provide better long-term options by jumpstarting the economic growth that people (of Africa) so desperately need, as he states 'more efficient agriculture will in general be better for the environment' and that 'the ability of civil society to assert itself will be much greater when people are prosperous and well fed'. But are there processes available to these communities to consider and achieve balance in such complex decisions? Possibly not, when we consider the track record of well-resourced first-world economies like Australia in dealing with similar instances, such as the ongoing debate over Murray-Darling Basin water allocation or Great Barrier Reef water quality.

Much of the developing world's agricultural research is directed towards improving production, and currently 'few plans for promoting sustainability have specifically built in means of either adapting to climate change or promoting adaptive capacity' (Pachauri 2011, p. 100). While a production focus of agricultural research for developing countries is appropriate, improvements in governance systems and the adaptive capacity of farmers are also required, and Australian agriculture is well placed to develop and extend practices that could be applicable and of benefit to other tropical regions; but Australia will need to *get it right* first. This is an opportunity for Australian research to deliver global benefit.

An important step in achieving this will be addressing Australia's skills loss and declining interest in agriculture as a career, as in recent years many Australian universities have either closed or merged their agricultural faculties to compensate for a shortfall in students, indicative of the broader negative community view towards commercial agriculture in Australia (Keogh 2009). Along with this demise, there continues to be a lack of clarity and certainty of farmer's property rights conferred by legislation, particularly for pastoral lease arrangements. Approaches vary across jurisdictions about non-pastoral land uses on leased land, but in general they are treated as special cases within the legislation. This lack of formal recognition reflects the continued narrow and prescriptive nature of pastoral lease arrangements (Productivity Commission 2002) and restricts the ability of farmers to pursue innovation. Rather than attempting to fix such issues individually, they would be better addressed through governance systems that recognise and value farmers' interdependence with society.

I mention these issues because the 2015 *Developing Northern Australia Whitepaper* provides a strategic opportunity for a growing Northern Australian agricultural sector to better prepare itself for the natural, social and economic opportunities and pitfalls it will encounter while embracing the principles of sustainable agriculture and recognising and fostering existing tropical expertise. This approach is in alignment with the proposal by B. Walker et al. (2010) to feature resilience and transformability alongside productivity as major objectives of research because, as previously stated, literature suggests that resilient organisations are as much about thriving as surviving. The dynamic relationships between vulnerability, resilience, hazard impact, hazard change, adaptive capacity and social change in the context of climate change and disasters can inform approaches to planning and developing community-based approaches to adaptation (Cottrell et al. 2011).

In the same way that Cottrell and King (2008) emphasise the need to have an understanding of how people living in communities view risk to more effectively engage them in planning and mitigation for disasters, both agricultural industry individuals and organisations will need to be engaged, understood and empowered as part of longer-term planning processes. This will require adoption of the principles described by D. King (2010) to achieve climate change adaptation – flexible, local, stakeholder driven and involving all levels of government and institutions.

While there are multiple views and contestations around resilience in human development, Michael Ungar, codirector of the Resilience Research Centre at Canada's Dalhousie University, challenges what he refers to as the dominant ecological understanding of resilience in human development and proposes that rather than being an objective fact, resilience is 'the outcome from negotiations between individuals and their environment for the resources that define themselves as healthy amidst conditions collectively viewed as adverse' (Ungar 2004). Ungar proposes that a more qualitative approach to resilience research will assist in informing international development, and while this work was among youth, I propose it is equally applicable to Northern Australia's agricultural context, and that successful outcomes would have international application.

This is important, because a characteristic of today's world is increasing urbanisation and social connectedness (Zalasiewicz et al. 2011); and while improved communication technology enables people in remote areas to easily and immediately connect to the rest of the world, it would seem that the ability of these same regional and remote populations to influence city-orientated political decision-making continues to decline. Deliberate mechanisms to maintain an inclusive society are required.

2.8 Conclusion

Resilience might be a fashionable concept and a difficult one to describe accurately; but it is also a widely used and generally understood term capable of bringing disciplines together for a better understanding of social-ecological issues. It can become a platform to support the deliberation around strategies to negotiate and shape current and future sustainability and is a concept that can assist individuals to navigate and negotiate challenges in their effort to find a way to live in relative comfort, both in response to and outside of risk and disaster.

This chapter, through a review of the resilience literature, shows how the concept has been used in many different disciplines and contexts, and that resilience has many definitions. Resilience can be seen at the level of the individual farmer, of the community, of an industry or of governance systems; it may focus on *bouncing back* (after a disaster) or *bouncing forward* (in the context of personal growth, adaptation and transformation); and it can also be seen both as an attribute of an entity and as a process. This book contextualises resilience through the perspective and words of individual farmers, which requires the consideration of the interplay of resilience at different levels be considered.

The following chapter will provide an overview of Northern Australia's agricultural development, followed by a consideration of contextual factors affecting its agricultural future in Chap. 4, before combining them with this resilience discussion.

References

Aldwin, C. M. (2007). *Stress, coping, and development: An integrative perspective* (2nd ed.). New York: Guilford Press.

Barrett, C. B. (2013, July 8, 2013). *Chronic rural poverty and resilience: Some reflections and a research agenda.* Paper presented at the James Cook University SIS group meeting.

Boon, H., Cottrell, A., King, D., Stevenson, R., & Millar, J. (2012). Bronfenbrenner's bioecological theory for modelling community resilience to natural disasters. *Natural Hazards, 60*(2), 381–408. https://doi.org/10.1007/s11069-011-0021-4.

Brennan, M. A. (2008). Conceptualizing resiliency: An interactional perspective for community and youth development. *Child Care in Practice, 14*(1), 55–64. https://doi.org/10.1080/13575270701733732.

Brown, K. (2016). *Resilience, development and global change*. London: Routledge.

Cocklin, C., & Dibden, J. (2009). Systems in peril: Climate change, agriculture and biodiversity in Australia. *IOP Conference Series: Earth and Environmental Science, 8*, 012013. https://doi.org/10.1088/1755-1315/8/1/012013.

Cote, M., & Nightingale, A. J. (2012). Resilience thinking meets social theory: Situating social change in socio-ecological systems (SES) research. *Progress in Human Geography, 36*(4), 475–489. https://doi.org/10.1177/0309132511425708.

Cottrell, A., King, D. (2008). Understanding communities' needs for information and education *Risk Wise*. (41–44). UK: Tudor Rose.

Cottrell, A., King, D., Dale, A. (2011). *Planning for uncertainty: Disasters, social resilience and climate change.* Paper presented at the 3rd World Planning Schools Congress, Perth. http://eprints.jcu.edu.au/18771/.

Cristaudo, A. (2012). Season end nears, but big issues still on the boil. *Australian Canegrower, 12 November 2012.*

Dahlberg, K. (2013). The scientific dichotomy and the question of evidence. *International Journal of Qualitative Studies on Health and Well-being, 8*, 21846. https://doi.org/10.3402/qhw.v8i0.21846.

Dale, A., Vella, K., Cottrell, A., Pert, P., Stephenson, B., King, D., et al (2011). *Conceptualising, evaluating and reporting social resilience in vulnerable regional and remote communities facing climate change in tropical queensland. Marine and Tropical Sciences Research Facility (MTSRF) Transition Project Final Report.* Retrieved from Cairns: http://www.rrrc.org.au/publications/research_reports.html.

Dale, A., McKee, J., Vella, K., Potts, R. (2013). Carbon, biodiversity and regional natural resource planning: Towards high impact next generation plans. *Australian Planner*, 1–12. doi:https://doi.org/10.1080/07293682.2013.764908.

Darnhofer, I. (2014). Resilience and why it matters for farm management. *European Review of Agricultural Economics, 41*(3), 461–484. https://doi.org/10.1093/erae/jbu012.

Davoudi, S., Shaw, K., Haider, L. J., Quinlan, A. E., Peterson, G. D., Wilkinson, C., et al (2012). Resilience: A bridging concept or a dead end? "reframing" resilience: Challenges for planning theory and practice interacting traps: Resilience assessment of a pasture management system in northern Afghanistan urban resilience: What does it mean in planning practice? Resilience as a useful concept for climate change adaptation? The politics of resilience for planning: A cautionary note. *Planning Theory & Practice, 13*(2), 299–333. https://doi.org/10.1080/14649357.2012.677124.

Flint, C. G., & Luloff, A. E. (2005). Natural resource-based communities, risk, and disaster: An intersection of theories. *Society & Natural Resources, 18*(5), 399–412. https://doi.org/10.1080/08941920590924747.

Folke, C., Carpenter, S., Elmqvist, T., Gunderson, L., Holling, C. S., & Walker, B. (2002). Resilience and sustainable development: Building adaptive capacity in a world of transformations. *Ambio: A Journal of the Human Environment, 31*(5), 437–440. https://doi.org/10.1579/0044-7447-31.5.437.

Folke, C., Carpenter, S. R., Walker, B., S, M., Chapin, T., & Rockström, J. (2010). Resilience thinking: Integrating resilience, adaptability and transformability. *Ecology and Society, 15*(4), 20.

Gill, F. (2011). Responsible agents: Responsibility and the changing relationship between farmers and the state. *Rural Society, 20*(2), 128–141.

Gray, I., & Lawrence, G. (2001). Neoliberalism. Individualism and prospects for regional renewal. *Rural Society, 11*(3), 283–298.

Hegney, D., Buikstra, E., Baker, P., Rogers-Clark, C., Pearce S, Ross H., King, C, Watson-Luke, A. (2007). Individual resilience in rural people: A Queensland study, Australia. *Rural and Remote Health, 7*(620 (Online)).

Higgins, V., Dibden, J., & Cocklin, C. (2010). Adapting standards – the case of environmental management systenls in Australia. In V. Higgins (Ed.), *Calculating the social: Standards and the reconfiguration of governing* (pp. 167–184). Houndmills: Palgrave Macmillan.

Hodgkinson, K. (2014, 7 July). Stoic to the limits. *The Daily Examiner.* Retrieved from www.dailyexaminer.com.au/news/letters/2310264/.

IAASTD. (2008). *Towards Multifunctional Agriculture for Social, Environmental and Economic Sustainability.* Retrieved from http://www.unep.org/dewa/agassessment/docs/10505_Multi.pdf.

IAASTD. (2009a). *Agriculture at a crossroads IAASTD_Global report* (978–1–59726-539-3). Retrieved from http://www.unep.org/dewa/agassessment/reports/IAASTD/EN/Agriculture%20at%20a%20Crossroads_Global%20Report%20(English).pdf.

IAASTD. (2009b). *Agriculture at a crossroads_Synthesis report.* Retrieved from Washington, DC: http://www.unep.org/dewa/agassessment/reports/IAASTD/EN/Agriculture%20at%20a%20Crossroads_Synthesis%20Report%20(English).pdf.

Juech, C., & Martin-Breen, B. (2011). Foreword. In P. Martin-Breen & J. Anderies (Eds.), *Resilience: A Literature Review* (pp. 1–2). New York: The Rockefeller Foundation. Retrieved from http://www.rockefellerfoundation.org/blog/resilience-literature-review.

Kaplan, H. B. (1999). Toward an understanding of resilience: A critical review of definitions and models. In M. Glantz & J. Johnson (Eds.), *Resilience and development* (pp. 17–83). New York: Kluwer Academic.

Kaplan, H. B. (2009). The revenge of geography. *Foreign Policy, 172*, 96–105.

Keogh, M. (2009). Agriculture in northern Australia in the context of global food security challenges. *Farm Policy Journal, 6*(2), 35–43.

King, C. (2008). Community resilience and contemporary Agri-ecological systems: Reconnecting people and food, and people with people. *Systems Research and Behavioral Science, 25*(1), 111–124. https://doi.org/10.1002/sres.854.

King, D. (2010). Issues in identifying climate change adaptation within community hazard mitigation. *International Journal of Emergency Management, Vol. 7*(Number 3–4). doi:https://doi.org/10.1504/IJEM.2010.037011.

Landry, M. (2014, 22 October). Princess Anne endorses Queensland beef. *The Morning Bulletin*. Retrieved from www.themorningbulletin.com.au/news/princess-anne-endorses-queensland-beef/2427771/.

Latour, B. (2005). *Reassembling the social: An introduction to actor-network-theory*. Oxford: Oxford University Press.

Lawrence, G., Richards, C. A., & Cheshire, L. (2004). The environmental enigma: Why do producers professing stewardship continue to practice poor natural resource management? *Journal of Environmental Policy & Planning, 6*(3–4), 251–270. https://doi.org/10.1080/1523908042000344069.

Lawrence, G., Richards, C., & Lyons, K. (2013). Food security in Australia in an era of neoliberalism, productivism and climate change. *Journal of Rural Studies, 29*(0), 30–39. https://doi.org/10.1016/j.jrurstud.2011.12.005.

Lele, S. (1998). Resilience, sustainability environmentalism. *Environment and Development Economics, 3*(02), 249–254.

Malcolm, B. (2004). Where's the economics? The core discipline of farm management has gone missing! *Australian Journal of Agricultural and Resource Economics, 48*(3), 395–417. https://doi.org/10.1111/j.1467-8489.2004.00262.x.

Martin-Breen, P., & Anderies, J. M. (2011). *Resilience: A literature review*. New York: The Rockefeller Foundation. Retrieved from www.rockefellerfoundation.org/blog/resilience-literature-review.

Masten, A. S., & Obradovic, J. (2006). Competence and resilience in development. *Annals of the New York Academy of Sciences, 1094*(1), 13–27. https://doi.org/10.1196/annals.1376.003.

Masten, A. S., & Obradovic, J. (2008). Disaster preparation and recovery: Lessons from research on resilience in human development. *Ecology and Society, 13*(1), 9. (Online).

McFarland, A. (2013). Careers. Retrieved from http://ausagventures.com/alix-mcfarland-regional-service-manager-for-nsw-farmers-federation/.

McManus, P., Walmsley, J., Argent, N., Baum, S., Bourke, L., Martin, J., et al (2012). Rural community and rural resilience: What is important to farmers in keeping their country towns alive? *Journal of Rural Studies, 28*(1), 20–29. doi:https://doi.org/10.1016/j.jrurstud.2011.09.003.

McNeely, J. A., & Scheer, S. J. (2003). *Ecoagriculture. Strategies to feed the world and save wild biodiversity*. Washinton, DC: Island Press.

Neales, S. (2015, January 3–4). Outback's big dry pushes farmers to brink. *Weekend Australian*, p. 13.

Nelson, D. R., Adger, W. N., & Brown, K. (2007). Adaptation to environmental change: Contributions of a resilience framework. *Annual Review of Environment and Resources, 32*(1), 395–419. https://doi.org/10.1146/annurev.energy.32.051807.090348.

Pachauri, R. K. (2011). *Striking the balance: Climate change, equity and sustainable development*. Retrieved from https://doi.org/10.1787/dcr-2011-en.

Polk, L. (1997). Toward a middle-range theory of resilience. *Advanced Nursing, 19*, 1–13.

Productivity Commission. (2002). *Pastoral leases and non-pastoral land use*. Retrieved from Canberra.

Reghezza-Zitt, M., Rufat, S., Djament-Tran, G., Le Blanc, A., & Lhomme, S. (2012). What resilience is not: Uses and abuses. *Cybergeo: European Journal of Geography [Online], 621*, 2–21. https://doi.org/10.4000/cybergeo.25554.

Rist, L., Felton, A., Nyström, M., Troell, M., Sponseller, R. A., Bengtsson, J., et al. (2014). Applying resilience thinking to production ecosystems. *Ecosphere, 5*(6), art73-art73. https://doi.org/10.1890/ES13-00330.1.

Sayer, J. (2010). Climate change implications for agricultural development and natural resources conservation in Africa. *Nature & Faune: Enhancing Natural Resources Management for Food Security in Africa, 25*(1), 17–20.

Skerratt, S. (2013). Enhancing the analysis of rural community resilience: Evidence from community land ownership. *Journal of Rural Studies, 31*(0), 36–46. https://doi.org/10.1016/j.jrurstud.2013.02.003.

Taylor, R. J. (1982). *The lost plantation: A history of the Australian tea industry*. Cairns: G.K. Bolton.

Taylor, B. M. (2010). Between argument and coercion: Social coordination in rural environmental governance. *Journal of Rural Studies, 26*(4), 383–393. https://doi.org/10.1016/j.jrurstud.2010.05.002.

Ungar, M. (2004). A constructionist discourse on resilience: Multiple contexts, multiple realities among at-risk children and youth. *Youth & Society, 35*(3), 341–365. https://doi.org/10.1177/0044118x03257030.

Ungar, M., Brown, M., Liebenberg, L., Othman, R., Kwong, W. M., Armstrong, M., & Gilgun, J. (2007). Unique pathways to resilience across cultures. *Adolescence, 42*(166), 287–310.

Walker B. (2014). *Evolution and resilience in the Anthropocene.* Paper presented at the Australian Academy of Science: Science at the Shine Dome 2014., Canberra.

Walker, B., & Salt, D. (2006). *Resilience thinking: Sustaining ecosystems and people in a changing world.* Washington, DC: Island Press.

Walker, B., Holling, C. S., Carpenter, S. R., & Kinzig, A. (2004). Resilience, adaptability and transformability in social–ecological systems. *Ecology and Society, 9*(2), 5.

Walker, B., Sayer, J., Andrew, N. L., & Campbell, B. (2010). Should enhanced resilience be an objective of natural resource management research for developing countries? *Crop Science, 50*(Supplement 1), S-10–S-19. https://doi.org/10.2135/cropsci2009.10.0565.

Wikipedia contributors. (2015 Aug. 26). Wicked problem. Retrieved 1 Sep. 2015a https://en.wikipedia.org/w/index.php?title=Wicked_problem&oldid=677921011.

Winkworth, G., Healy, C., Woodward, M., & Camilleri, P. (2009). Community capacity building: Learning from the 2003 Canberra bushfires. *Australian Journal of Emergency Management, 24*(2), 5–12.

Xu, L., & Marinova, D. (2013). Resilience thinking: A bibliometric analysis of socio-ecological research. *Scientometrics, 96*(3), 911–927. https://doi.org/10.1007/s11192-013-0957-0.

Zalasiewicz, J., Williams, M., Haywood, A., & Ellis, M. (2011). The Anthropocene: A new epoch of geological time. *Philosophical Transactions of the Royal Society, 369*, 835–841. https://doi.org/10.1098/rsta.2010.0339.

Zhou, H., Wang, J. a., Wan, J., & Jia, H. (2010). Resilience to natural hazards: A geographic perspective. *Natural Hazards, 53*(1), 21–41. https://doi.org/10.1007/s11069-009-9407-y.

Chapter 3
Agricultural Development in Northern Australia

You can make a small fortune in farming, if you start with a large one
Anon

3.1 Introduction

This chapter provides a history of both regional and agricultural development in Northern Australia, before progressing through a consideration of historic and contemporary challenges and drivers (regional, national and global). This information is then considered against existing policy paradigms and the perennial challenges of infrastructure provision in an extensive but lightly populated landscape.

3.2 The Historic Challenges of Northern Australian Agriculture

European explorers who considered Australia's First People as simple hunter-gatherers who relied on chance for survival and moulded their lives to the country where they lived might not have seen Australia as a farmed landscape, but work by Gammage (2011) clearly shows that pre-European Australia was in fact the 'biggest estate on Earth' – an artefact of tens of thousands of years of human intervention. As a consequence, the notion that Australia's 'natural' landscape can be maintained in its pre-European state simply by excluding or limiting contemporary land use does not stand up to scrutiny.

This immense paradigm shift has far-reaching consequences for contemporary Australia's attitudes towards both its First People and management of its landscape. Acceptance of Aboriginal Australia as a managed estate sustaining a stable human population is in stark contrast to early European settlement, where agriculture was

---

Keith Noble has contributed more to this chapter.

K. Noble et al., *Agriculture and Resilience in Australia's North*,
https://doi.org/10.1007/978-981-13-8355-7_3

basically subsistence – the colony was not self-sufficient in bread grains till the 1820s and relied on Tahitian pork till the 1830s (Heathcote 1994). While the number of Indigenous Australians living in the north now is much higher (14.3% of the population) than the national average (2.3%) (ABS 2015), the lifestyle and land use of Traditional Owners are now typified by relatively stable population centres and cessation of nomadic lifestyles. European settlement in the late nineteenth and early twentieth centuries was instrumental in this change, and it was the search for economic opportunity that drove the invasion of Aboriginal lands (Bottoms 2013).

From the mid-1850s, Australian agriculture underwent significant changes, partly the result of increasing local and export demand (driven by an influx of gold miners, repeal of protectionist British Corn Laws, higher European living standards, technological innovations), and partly the result of political policies on the process of land settlement itself (Heathcote 1975). A bright future was forecast for Northern Australia agriculture from the very beginning of European exploration: John McDouall Stuart, a Scottish-born explorer of inland Australia, remarked in his 1865 exploration account:

> I have no hesitation in saying the country I have discovered on and around the banks of the Adelaide River (near present day Darwin) is more favourable than any other part of the continent …. I feel confident that, in a few years, it will become one of the brightest gems in the British crown. (Stuart 1865, p. 6)

Aschmann (1977, p. 39) described the non-Indigenous development of Australia as:

> an initial investment of capital and introduction of people (sometimes involuntary), extensive pastoralism and limited subsistence farming, followed by mineral discoveries (often gold) that attracted enormous immigration. Extensive agriculture developed to feed these immigrants, often subsequently specialising in a commodity for export. Manufacturing and service industries developed to supply the established population, and the whole complex became economically self-sustaining;

though he thought that in the north, except along the east coast of Queensland, this sequence was interrupted at the agricultural stage not because of climatic or soil limitations but because, while transport was expensive, it was cheaper than local production – a cost-benefit outcome repeated in today's food mile debates.

Australia was on the winning end of many late nineteenth-/early twentieth-century innovations. Mechanisation (the petrol engine, refrigeration) combined with labour shortages (further driving the need to innovate) and large areas of land to allow Australia to supply cheap meat and butter to Europe and satisfy the growing demand for an improved working-class diet. Advances in pesticides, herbicides and fertilisers continued to increase productivity. In contrast, contemporary European farming systems were still based on labour-intensive methods and could not compete in open markets against mechanised industrial agriculture (Barr 2009). Following this period of rapid expansion, imperial preference deals between Britain and her colonies and dominions after World War I gave Australian agriculture a protected market. Food scarcities after global disruptions such as World War II, the Korean War, the Vietnam Wars and the Chinese Communist Revolution continued to provide export opportunities for undifferentiated commodities, as did extensive drought in the USA in the 1970s and 1980s and Middle East conflict in the 1990s.

The exporting of raw, or semi-processed, agricultural commodities such as wool, wheat, meat and sugar was the key to the early success of rural Australia, and continues to be the case to this day.

The first Commonwealth Administrator of the Northern Territory, Dr. Gilruth, placed great hope in the pastoral industry to develop the economy, and in 1914 the (British-owned) Vestey group of companies were allowed to build and operate the Darwin meat works, which in turn facilitated their control of vast pastoral leases. The meat works development doubled Darwin's population but, plagued by industrial action and failure to complete the rail line to the Katherine River, proved a dismal failure. They did not open till 1917, and closed in 1920 (Carment 1996).

World War II catalysed strategic concerns about the North Australia's 'emptiness' and highlighted opportunities for development. This resulted in a government 'imbued with a newly-forged nationalism and readiness to engineer the future' establishing the North Australian Development Committee (NADC) in 1946, charged with 'investigating the region's pastoral, agricultural, mining, forestry, marine, fuel and power, and processing and manufacturing industries; and to guide systematic development of these industries' (Garnett et al. 2008, p. vi).

Twenty years on, Dr. 'Nugget' Coombs, a NADC commissioner and long-time advocate for Northern Australia, said in his opening address to the first annual seminar of the Northern Australia Research Unit (NARU) in Darwin, August 1977, that:

> the optimism at the time and the prevailing views … that growth was a good thing, that it could be achieved primarily by seeking to impose on the North a pattern of productive activity and a way of life essentially European in its origin and substantially European in its relevance. There was little attempt to envisage the gradual emergence of a more humanized environment capable of self-perpetuation, providing a context for a more rewarding life for those who already lived within it. (Bauer 1977, pp. 8–9)

It is significant that this conference was titled 'Cropping in Northern Australia: Anatomy of Success and Failure': achievement of a productive landscape was the intent, and this could only be measured as a success or a failure. The proceedings of this conference contain an analysis by Fisher et al. (1977) of six large-scale agricultural developments (including the original Ord River Irrigation Project, Territory Rice at Humpty Doo, Tipperary Land Corporation and Lakeland Downs in Queensland), which showed that every one of them failed to achieve their stated objectives.

A particularly pertinent comment was made by Mollah (1980) in his retrospective analysis of the cropping development at Tipperary Station: in 1967, encouraged by a worldwide beef shortage, tax concessions to encourage investment, and the first stages of the Ord River development, the Tipperary Land Corporation (registered in Texas, USA) announced the biggest agricultural project yet attempted in Australia, with 'American know-how' and $20 million to establish a farming community of 15,000 people, producing 300,000 tons of grain sorghum annually and high-quality beef cattle. These great expectations were never realised, and farming was all but abandoned after 3 years, and the station sold back to Australian interests. Mollah commented (p. 156) that 'Pioneering developments in the North had no room for those who doubted their own ability but, from the outcome at Tipperary, it is equally clear that confidence is no substitute for knowledge and experience.'

Cook (2009) has summarised five historic pushes for Northern Territory development in which government-led research efforts assumed that once the science was in place, agricultural development would follow, and concluded the aspirational drivers for these initiatives related as much to addressing the perceived risks of Australia's 'empty North' as a genuine commitment to agricultural growth. Even in the more climatically hospitable and richer soils of Queensland's Atherton Tablelands, Gilmore (2005) considered government-sponsored agriculture was more a means of closer settlement and strategic defence than a food-producing venture; and that the maize, dairy and tobacco industries so established foundered when Australian governments realigned the economy according to neo-liberal principles.

Bauer (1985, p. 12) gave three reasons for this failure of large-scale agriculture in Northern Australia: (1) distance, (2) ignorance of the physical environment and (3) a reprehensible aversion to learning by experience. While these factors are neither exclusive to agricultural endeavours nor Northern Australia, item three is worthy of particular consideration in the contemporary enthusiasm for expanding Northern Australian agriculture.

3.3 Contemporary Challenges to North Australian Agricultural Aspirations

A viable and diverse agricultural industry operates in Northern Australia, with established production of beef, dairy, corn, sorghum, peanuts, avocadoes, mangoes, nuts, sugar and a myriad of fruits and vegetables as well as plantation timber. The emergence of successful new crops and cropping areas continues, such as chia[1] and extensive plantations of sandalwood around Kununurra (Western Australia) and the Burdekin (Queensland), while more speculative proposals continue to 'pop up', such as the proposed (and now lapsed) $2 billion Integrated Food and Energy Development (IFED) project on Queensland's Gilbert River (IFED 2014) and Stanbroke Pastoral Company's proposed $200 million irrigated cotton project in Queensland's Gulf Country (Crothers (2015) – evidence of the willingness of entrepreneurial growers (corporate and individual) to try new ventures.

Agriculture is an established and major contributor to Northern Australia's economy: in 2008/2009 direct primary industry turnover for Far North Queensland was estimated at over $1.7 billion, with direct employment of about 9000 people (RDA 2011). Australia is the world's largest exporter of sheep and cattle, and 80% of exported cattle are from the north. This trade was valued at $416 million in 2006–2007 (Gray 2009) and $1.4 billion in 2017 (though well short of the $22.3 billion projected for 2015–2016 (ABARES 2015)). And in a repeat of history, a new Darwin abattoir with eventual processing capacity of 220,000 head per annum (AACo 2012) opened in 2014, but ceased operations in 2018, citing lack of supply.

[1] *Salvia hispanica*, commonly known as chia, is a species of flowering plant in the mint family Lamiaceae, native to central and southern Mexico and Guatemala.

Australia's combined farm and livestock production gross value was forecast to reach $86.2 billion in 2015–2016 (ABARES 2015) but peaked at $56 billion. Despite this forecast discrepancy, Australian agriculture is estimated to contribute to the diets of 60 million people annually. While total Australian agricultural output represents only 1% of global production, Australia is the fourth largest net agricultural exporter in the world behind Brazil, Argentina and the Netherlands – well above nations such as China and the USA that have enormous domestic agricultural sectors but in net terms import just as much as they export (Keogh 2009). When the outcomes of agricultural research are also considered, Australia contributes to the diets of 400 million people worldwide (Prasad and Langridge 2012).

Most agricultural production is traded through efficient supply chains as bulk commodities, leaving little opportunity for an individual farmer to influence the price. Farmers are price-takers, so they must focus on maximising production while reducing operational costs (Hughes 2014). Keogh (2014) describes three factors that optimise Australian agriculture's international comparative advantage and provide opportunities for growth:

1. The flexibility of the farm business operating environment (achieved largely as a consequence of deregulation and the lowest level of OECD subsidies)
2. The technical and business management skills of operators
3. The open-minded and innovative approach of Australian farmers to new technology, illustrated by the rapid adoption of minimum tillage, digital farming technologies, advanced livestock genetics, precision agriculture and irrigation, electronic livestock identification, remote sensing and use of computerised decision support tools

The Australian Government's 2012 White Paper 'Australia in the Asian Century' describes opportunities for agriculture to benefit through forecast Asian growth, which Keogh (2012a, p. iv) interprets as a signal that policy-makers may now 'see opportunities for agriculture rather than as a sunset industry' and argues that Australian superannuation funds should join overseas pension funds in investing in agriculture (2012b). Other publications that also set out a strong future for Australian agriculture and its supply chains include the National Farmers' Federation (2012) *Blueprint for Australian Agriculture 2013–2020*, Australia's 2013 'National Food Plan: our food future' White Paper, the 'Agricultural Competitiveness White Paper' (2015a) and the 'Our North, Our Future: White Paper on Developing Northern Australia' (Australian Government 2015b).

Given that around 40% of Australia's land mass is north of the equator but less than 5% of the nation's population live there (ABS 2015), industry excitement about the potential for growth is understandable; but what are the aspirations of both northern resident and the broader Australian community? Not everyone wants to live in a city, and the continuing appeal of the frontier ethos as personified by Mary Durack's *Kings in Grass Castles* (1968) should not be dismissed, as the lure of big horizons and future opportunities continues to excite individuals and the nation alike (Bendle 2013), even though *Terra nullius*[2] never existed.

Generally though, regional residents desire equitable access to a similar suite of opportunities and support available elsewhere, and in particular a future for their children and access to services for an ageing population, and so it is worth noting that while absolute numbers have risen, the approximately 5% of the national population residing in the north (an overwhelming majority of whom live on the Queensland coast between Cairns and Rockhampton) is the same percentage and population distribution as in 1975, when Australia's first Department of Northern Development came to its ineffectual end, as well as 70 years earlier when President Roosevelt delivered his warning against leaving the north empty (McGregor 2016, p. 241).

The tyrannies of distance have been significantly addressed since Bauer's time, particularly through the mining boom of the past decade, with improvements in road and rail links, port infrastructure, communication systems and employment opportunities; albeit many of these being on a *fly-in fly-out* basis. Both agriculture and export-orientated mining have strong projected growth in Northern Australia's short- to medium-term future, and the obvious synergies of shared infrastructure are recognised (Owens 2013). Historically though, the internal distribution of costs and benefits from mining within host regions transitioning from agricultural economies has been limited (Hoath and Pavez 2013).

While Indigenous Australians' role in the past and present management of Australia's landscapes now appears obvious to most contemporary Australians, this has not been an easy transition, riven initially by uncertainty and community division over the implications of the 1992 Mabo High Court decision and the subsequent Commonwealth Government's Native Title Act 1993 impacts on private and leasehold tenure. Today, Indigenous Australians are majority landholders[3] of Northern Australia (CSIRO 2013; Moritz et al. 2013), and many are employed in community ranger and other land management programmes. These communities are actively addressing their social issues and considering alternate futures and meaningful employment opportunities for their growing communities (annual population growth is 2.1% compared with 1.6% for the national average (BITRE 2009)).

This is not to say that unanimity exists: consider the divided response to Queensland's Wild Rivers Act 2005[4] and contested development of the Kimberly (Western Australia) natural gas resource, where conservation agendas resulted in new alliances between Traditional Owners and graziers in the former and overt divisions between Traditional Owner communities in the latter, with Aboriginal leaders stating that Green Group's determination to maintain 'wilderness' areas distant

---

[2]A Latin expression deriving from Roman law meaning 'nobody's land', which is used in international law to describe territory which has never been subject to the sovereignty of any state, or over which any prior sovereign has expressly or implicitly relinquished sovereignty. Sovereignty over territory which is terra nullius may be acquired through occupation, though in some cases doing so would violate an international law or treaty (Wikipedia: The Free Encyclopedia 2015, July 21).

[3]While some land is held as unencumbered freehold, native title is not freehold and a restricted form of tenure.

[4]Act No. 42 of 2005: An act to provide for the preservation of the natural values of wild rivers and for related purposes (Queensland Government 2005); repealed in 2014.

from the comfortable suburbs in which most of their supporters live deprives Indigenous people of the economic opportunities they need to end poverty and social marginalisation (O'Faircheallaigh 2011). On the international stage, Pieck and Moog (2009) describe similar schisms in the iconic Amazon eco-Indigenous alliance due to Indigenous people never becoming a central part of the agenda of large conservation organisations and their subsequent failure to live up to the discursive (and promotional) assurances they made to Indigenous people. Cruz (2010, p. 421) warns that 'publishing information on a webpage does not make it more accessible to members of a local community, but rather allows that knowledge to escape local control and be used by anyone'.

The driver for many environmental campaigns has been concerns over resource exploitation. However, mining is increasingly providing opportunities for Aboriginal business and employment in Northern Australia: Fortescue Metals, an iron ore miner in the Pilbara area of Western Australia, have invested $1 billion with Aboriginal businesses since 2010 and employ 1000 Aboriginal people (Power 2013). This experience contrasts sharply with historical agricultural industry engagement with Aboriginal people, characterised initially by dispossession and persecution, before moving through exploitation to widespread disengagement. Today, the Far North Queensland banana industry is heavily dependent on international backpacker labour, while aboriginal communities in the same region experience chronic underemployment.

In February 2011, tropical cyclone Yasi cut a swathe through Far North Queensland's natural and production environments, compounding the impacts from cyclone Larry 5 years before. Unlike cyclone Larry, Yasi occurred in the aftermath of the global financial crisis and in the context of a high Australian dollar and a 'summer of disasters' throughout eastern Australia. Agricultural industry impact was widespread but highly variable. The banana industry received extensive media coverage with 90% destruction of the Australian crop, but within 10 months was back in full production and facing chronic market oversupply and resultant low prices. The tropical fruit industry however was dealt a crippling blow that it still has not recovered from. Sugar cane, cattle, dairy and other tree and horticulture crops were also affected to varying degrees relative to their geographical relation to the cyclone's path, and recovery times from Yasi did not mirror those from Larry 5 years earlier (personal experience and observation). How can an agricultural industry expansion be contemplated when the incidence of such disasters is predicted to increase as a consequence of climate change?

A factor made clear by two high-intensity cyclones impacting the same area within 5 years was the interdependence of industries across scale and commodities: without regular banana transport south, freight costs escalated to the point government subsidy was required to ensure affordable delivery of much-needed building materials, but farmers capable of sending product south also required freight assistance; the disappearance of backpacker employment impacted on local accommodation and tourism businesses; and dairy cows could not be milked without electricity.

Additionally, while cyclones are stand-alone and geographically defined events, their impact manifests in the convoluted environment of world markets. As an example, the greatest concern of the Australian Banana Growers Council post-Yasi was that the lack of supply to supermarkets would result in the importation of bananas from the Philippines (Australian Food News 2011), an event that would have more significant and persistent industry impact than one cyclone, and again demonstrating that disasters are not limited to natural events.

3.4 National Drivers for Growth

Contemporary drivers for developing Northern Australia appear little changed from those of earlier times: 'untapped promise, abundant resources and talented people … closest connection with our trading partners … a strong north means a strong nation' (Australian Government 2015b, p. 1), though the 2015 Developing Northern Australia White Paper does describe government's role as facilitator rather than leader of growth, and commits to trialling and testing new policies 'rather than just relying on the lessons of the south' (p. 10).

The twenty-first century has seen a focus on contemporary Indigenous management of the Northern Australian natural landscapes for the provision of environmental services (Cook et al. 2012), but neither the failure of past agricultural endeavours nor the emerging recognition of Indigenous environmental stewardship has stopped the ongoing speculation about Northern Australia's opportunities for further agricultural development. The twenty-first-century drought in Southern Australia (compounded by over-allocation of irrigation water) fed this debate to the extent that in 2007 then Prime Minister John Howard established, as part of his plan for water security, a Northern Australian Land and Water (NALW) Taskforce to:

> examine the potential for further land and water development in Northern Australia, with particular emphasis on the identification of the capacity of the north to play a role in future agricultural development. (Garnett et al. 2008, p. vii)

While some industry sectors considered the projections and assumptions in the NALW report conservative (Maher 2011), the report clearly states that the future of the north should not be limited to pastoralism and/or irrigated agriculture and that decision-making should be based on a thorough and balanced assessment of the economic, social and environmental implications. The commitment to such a decision-making context is given in the *Developing Northern Australia Whitepaper* which, assisted by ongoing technological developments, would address the first two of Bauer's concerns, but what of 'the reprehensible aversion to learning by experience'?

McLean and Gray (2012, p. 196) promote the 'thinking use of history' as a mechanism for reinterpretation of the premises of major policy decisions. Lessons from the past highlight the need for precautionary action to act in the anticipation of change and in the face of uncertainty, to work at multiple scales, to develop institutions which will adapt and to incorporate learning. Without this, the potential for repeating past failure remains, particularly if the underlying policy paradigms are ill-conceived or flawed.

3.5 Global Drivers for Growth

The burgeoning world population of the twentieth century combined with political instability to bring (via improved communications and the international media) the reality of famine to the living rooms of the developed world in the 1970s and 1980s. These images, along with NASA photos of Earth from space, brought the realisation to many people that globalisation meant more than cheap air travel and Americanisation of the Australian language. Globalisation meant awareness of national inequities, and a sense of obligation (at least in some) to address these. The political geographer Kaplan (2006) describes globalisation as a cultural and economic phenomenon, and not a system of international security. Kaplan (2009) also postulates that 'like rifts in the Earth's crust that produce physical instability [there are areas in the world] more prone to conflicts than others' (p. 102) and that 'rather than eliminating the relevance of geography, globalisation is reinforcing it' (p. 98). Kaplan reasons that since the world has realised international relations are ruled by a sadder, more limited reality than the one governing domestic affairs (of his home, the USA), the central question of foreign affairs is now 'Who can do what for whom?' In response, the developed world's population gave a clear signal that starvation was something it could, and would, do something about.

Global food marketing is a dynamic process open to manipulation by individual countries through measures such as import controls, export subsidies and price guarantees. The international deregulation of agriculture, including market reform and the abolition of subsidies and protection, has been a major part of globalisation. Restructuring of agriculture is a primary outcome of deregulation and is itself a major cause of impact on rural communities, and while globalisation may bring benefits, it also has adverse impacts which are not seriously considered by many economic analysts (Vanclay 2003). Aggregate demand for food in developed countries increases only slowly (McCrone 1962, p. 121), and the income elasticity of demand for food declines with economic growth such that as per capita incomes increase, the proportion of income spent on food declines (Engel's Law[5]). In contrast, higher-yielding crop varieties, more prolific livestock strains, improved husbandry and mechanisation all lower unit production costs and increase agricultural productivity, resulting in a treadmill where innovative farmers apply revenue-increasing or production cost-decreasing technology which other farmers subsequently adopt, so production increases while prices decrease. Interpreted in this way, technology is no longer a means but a cause of agricultural adjustment, ably demonstrated by the 55% increase in US total farm output between 1940 and 1960, while capital inputs only increased by 5% (Bowler 1979). By the 1980s, this treadmill of production combined with increasing awareness of environmental and heritage issues, and transnational adoption of neo-liberal principles, to force the developed nation's farm sector transition from a productivism regime to one of post-productivism in the later stages of the twentieth century (Bindraban & Rabbinge 2012).

[5] The law was named after the statistician Ernst Engel (1821–1896).

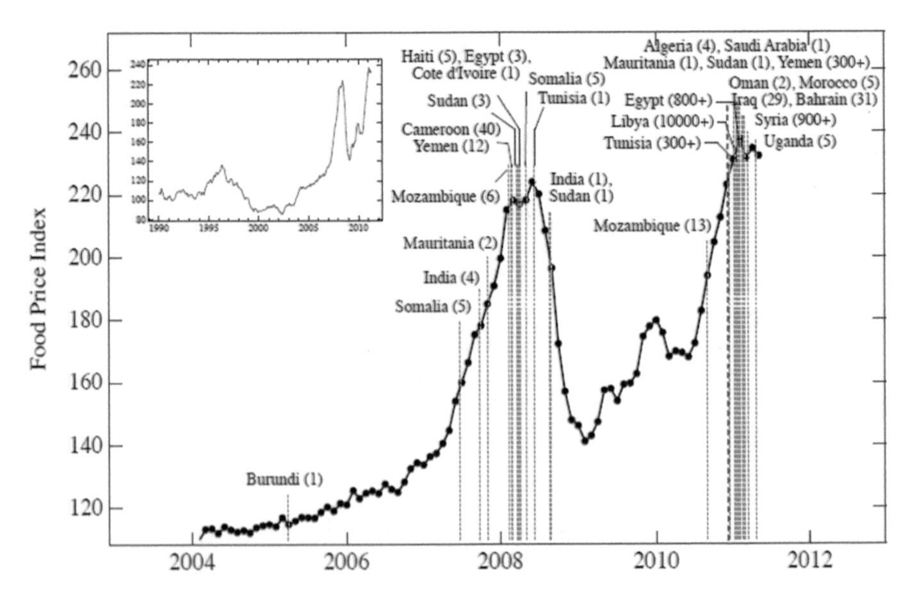

Fig. 3.1 Time dependence of FAO Food Price Index, January 2004 to May 2011. Red dashed vertical lines correspond to beginning dates of 'food riots' and protests associated with the major recent unrest in North Africa and the Middle East. The overall death toll is reported in parentheses. Blue vertical line indicates the date, 13 December 2010, on which the authors submitted a report to the US government, warning of the link between food prices, social unrest and political instability. Inset shows FAO Food Price Index from 1990 to 2011. (Source: Lagi et al. (2011). *The Food Crises and Political Instability in North Africa and the Middle East.* Social Science Research Network)

The world in 2050 is expected to be not only hungrier[6] and wealthier[7], but also fussier, with consumers empowered by information expecting food to be nothing less than healthy, nutritious, clean, green and ethically produced (Hajkowicz and Eady 2015). International interest in food security was focused by the 2007–2008 spike in world food prices (see Fig. 3.1) as a consequence of a shift from food to biofuel production (Fraser 2008). It has been retrospectively argued that the crisis was actually triggered by a combination of short-term factors and longer-term trends, including a series of extreme weather events, low global stock levels, the use of food crops for biofuels, rising energy prices, export bans and increased financial speculation, along with structural problems rooted in global resource limits (see Maye and Kirwan (2013)). But politicians and policy-makers were left in no doubt as to the potential threats to food security, the increasing interdependence of agri-food systems and the political and social importance of affordable food (Ambler-Edwards et al. 2009; Lagi et al. 2011).

The United Nations Food and Agriculture Organisation World Food Security Summit in Rome, June 2008, helped establish a consensus that food security was a

[6] 2.4 billion more people on Earth requiring 60–70% more food than currently available.

[7] Over 1 billion people in Asia alone will move out of poverty as average incomes almost quadruple by 2060.

key *master frame* of twenty-first-century public policy (Mooney and Hunt 2009) and that the risks to food security also included slower-onset, more diffuse perturbations such as global climate change (Ericksen et al. (2009), Misselhorn et al. (2012)).

However, as with most complex problems, the opportunity for a simple solution quickly fades under scrutiny. In 2015 there were approximately 795 million people undernourished globally, down 167 million over the previous decade and 216 million less than in 1990–1992 (Food & Agriculture Organisation 2015). Yet almost 30% of people globally are now estimated to be either obese or overweight – a staggering 2.1 billion (Ng et al. 2014) – making the Anthropocene[8] the first time in human history that more people have to choose to eat less to improve their health, than those who need to eat more.

But starvation is 'the characteristic of some people not having enough food to eat, not the characteristic of there being not enough food to eat' (Sen (1981, p. 1) in Mooney and Hunt (2009)) which means that hunger as a component of food security is more a question of access than availability, and research by West et al. (2014) indicates opportunities exist to improve both global food security and environmental sustainability for another three billion people through a relatively small set of places and actions: principally yield increase and waste reduction from current production areas, none of which include Australia.

This brings into question the fundamental approach by global agricultural science and technology for more than a century that

> the adoption of new technologies by entrepreneurial producers willing to take enough risk in the global marketplace to increase their productivity in the pursuit of profit, with the alleviation of hunger presumably following as a latent function (Mooney and Hunt 2009, p. 487)

because, managed efficiently, the world has enough ability to feed growing populations. In the future, agricultural development 'will have to serve the multi-targeted demands on agricultural production systems (food and other land-based products, conservation, resource protection, climate mitigation)' (Müller and Lotze-Campen 2012, p. 92), and Maye and Kirwan's editorial in the 2013 special issue on food security warns of the risks to agriculture's other outputs such as nature conservation and water management from pressures to produce food and energy. In the same issue, Allen (2013, p. 137) warns that 'solutions that move control farther from the ability of people in their everyday lives should be subjected to particular scrutiny' and that productivist goals of doubling food output could exacerbate – not solve – problems associated with food insecurity such as energy costs, climate change and food of low nutritional quality, and suggests that food security is a collective problem requiring a social solution. Innovative outcomes are coming from this debate though, an example being *Project Catalyst* – a Coca-Cola/WWF/Australian Government/NRM Regional Bodies/sugar industry partnership aimed at reducing the environmental impacts of sugar production on Great Barrier Reef water quality

---

[8] The Anthropocene is a proposed epoch that begins when human activities started to have a significant global impact on Earth's geology and ecosystems. Neither the International Commission on Stratigraphy nor the International Union of Geological Sciences is yet to officially approve the term as a recognised subdivision of geological time.

through innovative farming practices (WWF 2013) and along the way substantiating Coca-Cola's social licence to source sugar from a politically stable but environmentally sensitive area (Cocco 2013).

The Australian International Food Security Centre was established in 2012 to consider Australia's role in feeding an extra two or three billion people without irretrievably damaging the planet, and focuses on Australia's role in research and extension (Blight 2012). The Department of Foreign Affairs and Trade Secretary Peter Varghese told a panel on capitalising on the Asian Century that 'in a globalised world … success lies as much in our degree of internationalisation as in domestic factors' (CEDA 2013). But Australia's agricultural history differs from Europe and the USA, being export-orientated around demands for food and raw materials, more often driven by foreign investment and resulting in a lack of (or unwillingness to) finance value-adding or processing capacity (Burch et al. 1999). This has locked production into low value, unprocessed, competitive commodities vulnerable to free market price fluctuations and exacerbated by increasing application of neo-liberal policies (Lawrence et al. 2013). Australia's agricultural sector is the second least protected (subsidised) in the OECD[9] (O'Meagher 2005). So despite recognition of the need to value-add, most Australian food exports continue to be commodities processed to the minimum necessary for stability and transport, through supply chains fragmented and dominated by overseas interests (Ball 2012).

The tropics is home to 40% of the world's population and 55% of children under the age of 5, and by 2050 these figures are predicted to grow to 50% and 60%, respectively, while the tropical economy is growing 20% faster than the rest of the world (JCU 2014). More than half the world's middle class is expected to be Asian by 2050, and this wealthier population will require more food and prefer food of higher value, particularly meat and processed foods (Australian Government 2012; Hajkowicz and Eady 2015). However, while Northern Australia is one of the largest land masses in the tropical region, it remains the most sparsely populated, a situation that continues to raise national defence concerns in some sectors.

The interplay of issues related to northern agricultural expansion includes water resource management, climate change, disaster management, environmental impacts (particularly on, but not limited to, the Great Barrier Reef), native title and other tenure issues, the future of regional communities, foreign ownership of land and production units,[10] the economics of developing new irrigation areas and who will pay for the infrastructure. The role Northern Australia will play in future world food security is therefore politically and socially contentious in addition to the structural issues raised by Bauer in 1977.

[9] The Organisation for Economic Co-operation and Development was established in 1961 and provides a forum in which governments of 34 countries can work together to share experiences and seek solutions to common problems (OECD 2015).

[10] While foreign ownership is an emotional issue, the Australian Bureau of Statistics indicate that in 2013 just under 99% of Australian farm businesses and just under 90% of farmland is fully Australian owned (ABS 2014).

3.6 Agricultural Policy Paradigms

While many societies, including Australia, have their underlying value system expressed in a constitution, usually these values are translated into more tangible beliefs of how the economy or society should be organised and expressed as policy goals. Or, as Martin et al. (2011, p. 1) prefer, a social contract[11] that

> underpins core elements of the Australian Federation in respect to the national view of assuring a sustainable future for those areas that might not otherwise share in the distributed wealth of the nation.

Australian agricultural policy is therefore a component of a broader attempt to achieve shared aspirations or values in our society and cannot be limited to the agricultural sector. Agricultural policy is developed within the broader arenas of regional, national and global policy, all of which ultimately must deliver outcomes in relation to the economy, the community and the environment. Therefore, policy settings can, and do, vary over time. Habermas (2010, p. 191) understands this to be a dynamic process and that

> constitutional democracy [such as Australia's] depends not only on a routine system of checks and balances or procedural norms governing the making and implementation of law and so on, but on the active intervention of citizens into the political process, whose capability to "restore" the proper normative relationship among citizens, and between citizens and the state, rests on their ability either to foster or at least exploit a "crisis" situation that impairs the political system's normal operating parameters.

Such was the case in the early settlement of Australia. Although there were no real physical barriers to pastoral land use expansion once the Blue Mountains had been crossed by European settlers in 1813, there were considerable legal barriers until 1847. Initially, any occupation of the interior had been forbidden (but not necessarily enforced) by the British Government to concentrate energies and security of the small population closer to the coast. From the 1830s though, official opinion came to see pastoral land use as an effective form of pioneer settlement, in the tracks of which farmers could follow (Heathcote 1975). The system of leases adopted in 1847 was based on the premise that

> In return for occupation without ownership, he (the selector) has been granted use by lease of only a limited range of the land's resources. If he wishes to change his land use to agriculture, he was and still is usually required to give up his pastoral lease and purchase his land (p. 91)

– a premise that was to remain the feature of pastoral land use until the 1970s, after which some alternative land uses were considered for inclusion within a lease.

Beginning in the 1850s, agricultural land use was encouraged by colonial and state governments with the intention (firmly stated in all local parliamentary debates) to break the early squatter's land monopolies and 'socialise' this natural resource by

[11] A concept outlined by the Swiss-French philosopher Jean-Jacques Rousseau in 1762 of systems and processes in Western democracies that seek to ensure the equitable allocation of resources for basic infrastructure and services such as roads, health and education.

the establishment of a *yeoman* farming settlement. These 'closer settlement' policies were intended to lead to intensification of capital investment and consequent production, but through a combination of careless legal drafting, inefficient supervision of enforcement and blatant frauds, the legislation was less effective than hoped (Heathcote 1975, p. 109). However, in the sparsely settled north of Queensland, the (new) 1884 Land Bill resulted in constantly increasing rents without security of tenure (Bennett 1927) and 'returns so small it was possible to carry on only by producing great numbers [of cattle]' (p. 217) with concomitant over-exploitation of pasture and environmental degradation. Subsequent governments repealed some aspects of this Bill and provided subsidies for fertiliser use, incentives for tree clearing, funding for expansion into the drier areas of the continent and extension and education services that ensured Australian agriculture would take advantage of the latest technological and managerial innovations.

Governments also fostered the creation of grower boards and statutory marketing authorities, founded research institutes to develop and apply innovations and employed a cadre of extension officers to communicate the results of science to the farming population (Lawrence et al. 2004). In many instances these assistance measures also distorted resource use across farms and weakened farmer's incentives to find better ways of managing risk and improve productivity. Moreover, government assistance served to offset normal adjustment pressures, impeding ongoing structural change and preventing more efficient farmers from expanding their operations (E. Gray et al. 2014).

In part a consequence of, but in line with global trends, the 1980s and 1990s saw the adoption of *neo-liberalism* or *hard liberalism* by Australian governments, described as the most profound transformation of Australian public policy since World War II and one that fundamentally reworked a framework in place since federation (Western et al. 2007). This transformation was underwritten by two principles: (1) liberalism, the view that citizens are autonomous individual actors whose interests are best served when they are free from coercive government interventions into individual action (Yeatman 2000), and (2) marketisation – the belief that free markets are arenas which best enable individual autonomy and produce efficient economic outcomes (Marginson 1997).

In Australia, as elsewhere, the primary arguments for neo-liberalism were economic: that free markets are necessary for sustained economic, employment and income growth and were an inevitable consequence of an increasingly globalised world. Governments also privatised existing marketing arrangements for many commodities whereby farmers were obliged to sell their produce through a government monopoly (single desk) marketing board. After neo-liberal deregulation, farmers could access a range of marketing options and strategies including future contracts, but these new arrangements also presented difficulties for those with no experience of them. While some prospered, many struggled to understand the changes, and some farmers suffered large losses from the increase in risk (Vanclay 2003).

There is also widespread belief that in pursuing competition policy, governments deliberately preferenced the interests of consumers in cheap food products over the interests of producers in receiving a fair return for their labour and investment

(Edwards 2018). While the contributions of markets are central to the welfare of citizens and the environment, this should not leave the responsibility of ensuring a coherent social system to economic interests alone, and any 'policy strategy that seeks to increasingly shift the responsibility for administering society to business interests appears quite incomplete' (Hogan and Young 2013, p. 329).

The application of neo-liberal policies was across all sectors of the Australian economy, but Burch et al. (1999, p. 183) describe how when neo-liberal restructuring occurred in advanced economies such as Australia

> Globalization uniquely manifests itself in the transfer of control of existing (food) processing facilities and their reconstitution into a broader corporate plan. This has involved agri-food processing companies making decisions to cease production of a particular commodity, or to shift production to new, cheaper, sites within Australasia or further afield

a process which Vanclay (2003) refers to as *jurisdiction shopping*. Australia was particularly vulnerable to this scenario because its long history of export agriculture was organised around the demands for basic foodstuffs and raw materials, with only a small local market for food products and a minimal commitment to value-added manufacturing. This had made Australia peculiarly dependent upon one or two major markets for agricultural exports. Initially Britain was the major outlet for Australian produce, but once Britain joined the European Economic Community, Australia was forced into a radical reorientation to find new markets for wool, beef, wheat and the wide range of other products that it produced so cheaply (Aschmann 1977; Burch et al. 1999).

Today, Australian governments have removed almost all statutory marketing arrangements that characterised the pre-1990s agricultural marketing sector, and Keogh (2014) identifies this as a contributor to the flexibility of today's farm business operating environment, with farmers free to adjust enterprises and management systems in response to market signals. Keogh describes this as a major success factor, especially for broadacre farmers, and claims that 'farmers have no desire to return to a more regulated market environment' (p. 4).

Collits (2006) believes that most regional policy objectives in Australia can be reduced into two overarching goals: (1) decentralisation, with less regional out-migration, retention of young people in regions, greater economic diversity and improved employment opportunities, and (2) reduced interregional disparities. In 1965, Max Neutze observed that 'decentralisation of population has been the policy of most political parties since World War II. It has, so to speak, been everyone's policy but no one's programme' (p. v). By the late 1970s, Australian governments had moved away from an overt decentralisation objective and pursued the 'more familiar overseas policy concern of reducing disparities … even though the decentralisation urge remains strong in many regions' (Collits 2006, p. 2). The reasons for this change were complex and described by Collits (2011, p. 44) as a 'classic "wicked problem" [where] just about everything that has been thought of has already been tried … to little real effect'.

Centralisation was not just about economic rationalism or metro-centrism, tempting though these explanations might be. Rather, Collits (2011) suggests governments abandoned decentralisation for a complex and intertwined set of reasons which include:

- Serious difficulties with the idea of decentralisation itself
- The emergence of new regional problems because of changing regional conditions that overshadowed the old problem of metropolitan primacy
- The advent of new thinking about regional development and its drivers and new regional policy ideas more relevant to the emerging problems
- The ineffectiveness of interest groups supporting decentralisation
- Changing ideologies within government and changing policy priorities of governments that conflicted with balanced development
- Institutions and processes within government that were unsympathetic to decentralisation.

Around the world, the greater the importance of agriculture in an economy, the stronger has been the belief in the advantages of a rural life, with *farm people* considered to make a special contribution to political, economic and social stability (Hathaway 1969). The twentieth century saw increased urbanisation of many of the developed world's populations and an associated decline in the rural influence, to the extent that Self and Storing (1962, p. 226) observed that, 'the question is no longer whether the family farm will save democracy, but whether democracy is prepared to save the family farm'. Half a century later, this remains a publicly debated question without an agreed answer, complicated now by foreign farm ownership and the globalisation of agriculture.

Paradoxically, while farmers worldwide are renowned for their beliefs in independence and free enterprise, they have also been active in political processes to deliver a disproportionally large influence on policy-making. While Wilson (1977) could find no common reason across different countries for agriculture's political power, Bowler (1979) generalised three main factors which explained the power of agriculture in the political processes of many developed countries at that time:

1. Individuals in agriculture have formed active and vociferous *interest groups*,[12] usually apolitical, and able to form close relationships with government agencies and so circumvent established political channels.
2. Widespread acceptance that the social and economic problems of agriculture are different from other national problems, as agriculture is not the master of the physical environment it operates in, leading to establishment of separate Ministries for Agriculture which provided direct access to government through their own department (and which often became the sector's spokesperson in government decision-making).
3. Direct political voting power of farmers, farm workers, their families and the broader rural vote.

[12] *Interest group* embraces both groups hoping to obtain a material advantage based on a common objective and those held together by a common ideals or policies.

Increasing urbanisation has diluted this influence, though in the 1980s rural areas in many countries still tended to be proportionally over-represented in legislating bodies as constituency boundaries were not revised with sufficient regularity to counteract changes in population distribution or, in some cases, via a direct gerrymander.[13] These three points, but particularly the first, align with Habermas's conclusion that only politicians and administrators remain the ultimate arbiters of public good, because

> only their actions are governed by explicit and binding democratic procedures, which are understood to manifest themselves only in the relationship between the "strong" public sphere of representative legislatures, the political parties that occupy it, the administrative apparatus and the legal judiciary. Habermas (2010, p. 192)

Times continue to change, and there has been a continuing reduction in direct political influence by farmers; but the wellbeing, security and future of Australia's agricultural sector continue to figure prominently in public discussions. While not limited to questions of production, the broader Australian public is demonstrably interested and engaged at many levels on wide-ranging agricultural issues including foreign ownership of farm lands, animal welfare, food miles and provenance, agricultural impacts on the natural environment, mining impacts on the agricultural environment, food safety, food security, Australia's supermarket duopoly and its relationship with farmers, regional food processing capacity, drought and natural disaster impact on mental wellbeing in rural areas, social equity and service provision and, of course, our national image.

Australia might be one of the world's most urbanised nations, but outback sunsets, endless plains and new frontiers still feature strongly in the national identity (and tourism promotion), with 'a wistfulness about, and longing for, the days of decentralisation policy among the people in regional Australia' (Collits 2011, p. 42). Notwithstanding this sentiment, Sorensen (1993, p. 238) suggests that

> In general terms, Australia's market driven settlement system is well attuned to the nation's geography … there is, for good reason, no large city in the interior … it therefore seems eminently logical to have several large metropolises that are capable both of providing high order services and dealing with the rest of the world as equals, and to surround them with a range of small regional service centres and their tributaries. It is patently absurd to look to settlement systems in the quite different geographical environment of Europe and North America and claim that Australia is in some way deficient.

While the *Our North, Our Future: White Paper on Developing Northern Australia* (2015) includes strategies to support economic and population growth, there is no intent to develop a northern metropolis.

---

[13] A gerrymander is where electoral boundaries are redrawn in an unnatural way with the dominant intention of favouring one political party or grouping over its rivals. In Queensland, the Bjelke-Petersen government's 1972 redistributions occasionally had elements of "gerrymandering" in the strict sense, though their perceived unfairness had more to do with malapportionment whereby certain areas (normally rural) were simply granted more representation than their population would dictate if electorates contained equal numbers of voters.

3.7 The Importance of Infrastructure

The provision of infrastructure including transport, energy, water, communications and built human environments provides services essential for economic activities including Australia's domestic and international food supply chains (Nguyen et al. 2013). Distance, isolation and low population densities however have always been particular challenges for provision of such infrastructure to those who ventured away from south-east Australia, and their resolution has been a major hurdle which (it is popularly believed), if overcome, would facilitate more settlement, increase productivity and improve quality of life. The scale of the challenge is such that it is popularly seen as the role of government to address or, in other words, beyond the capacity of the affected individuals.

However, it also needs to be remembered that for some individuals the isolation and challenge of Northern Australia is part of the regional attraction, along with the sense of opportunity that accompanies new horizons. Notwithstanding this, low population numbers combined with tropical geographic and climatic constraints continue to present challenges for infrastructure provision. In an analysis of Commonwealth Government funding, Hull (2000) found that Northern Australia received considerably more than its pro-rata population share of grants[14] (5.3% of the population but 8.4% of grants). This could be attributed in part to higher maintenance and depreciation costs because of the physical environment's tropical nature, but other identified factors included socio-demographic composition; administrative scale; differential input costs for labour, accommodation and electricity; reduced urbanisation affecting service scale delivery; dispersion and isolation; and being uneconomic for the private sector to provide services.

The question of infrastructure provision is an important element of the 2015 White Paper on Developing Northern Australia, with specific provision of a $5 billion Northern Australia Infrastructure Facility (NAIF) to provide concessional loans for the construction of major infrastructure such as ports, roads, rail, pipelines and electricity and water supply. This recognition of the critical role of infrastructure in northern development is illustrative of the importance of understanding the infrastructure provision context, past and present, on agricultural development and farmers, and this section is devoted to further explanation of these influencing factors.

Before doing so though, it is important to point out that despite the $1.1 trillion forecast to be spent on critical infrastructure between now and 2050 and despite Australia anticipated to become a signatory to *The Sendai Framework for Disaster Risk Reduction 2015–2030*[15] (UNISDR 2015), there is currently no formal requirement to consider resilience to disasters when making decisions about building infrastructure (Deloitte Access Economics 2016).

[14] This analysis excluded funding to Australia's northern external territories – the Cocos (Keeling) Islands and Christmas Island.

[15] The Sendai Framework includes preparedness to 'Build Back Better'.

3.7.1 Freight and Transport Infrastructure

As a lightly populated continent, Australia's economy is both dependent on and sensitive to freight and transport costs. The Australian Government 2011 National Land Freight Strategy describes how many freight systems have their origins in the pre-federation colonies, so any national freight network discussion necessarily has historical overtones. For example, federation-era transport aspirations were for standardised rail gauge, and though there have been incremental changes to improve connectivity, the regulatory break of rail gauge between states still exists.

All levels of government are engaged with freight, and transport policies often reflect principles that governments want pursued across sectors. Various maps and definitions of a national transport network have been produced (Auslink/Nation Building, the National Highways scheme, the Designated Interstate Rail Network, the Australian Rail Track Corporation network), but none depict a national land freight network. In most respects, policy initiatives have been directed at providing infrastructure to be used by transport rather than identifying freight as a separate target, and many transport plans relate primarily to the application of government funding.

3.7.1.1 Sea Freight

Sea freight provided transport from the very beginning of European exploration and settlement; and as a nation dependent on maritime trade, Australia's ports remain an important gateway for agriculture, with operational ports around Australia providing transport logistics for 42.4 million tonnes of agricultural production in 2012/2013, comprised of wool, cotton, sugar, grain, livestock and timber products. Total port throughput was 1.1 billion tonnes, the balance being predominately mineral exports (Port Master Planning 2014).

The resources boom of Australia's twenty-first century resulted in rapid expansion of some port facilities and associated infrastructure, particularly in Northern Australia, and Australia's bulk commodity exports and metropolitan container imports are both expected to double in size every 10 years, resulting in some regional bulk export ports now facing significant infrastructure challenges (COAG 2011). In 2018 the Port of Townsville, North Australia's largest port, received $193 million for a channel capacity upgrade – widening the existing channel from 92 to 180 m to accommodate the significantly bigger ships now used for sea freight (ships longer than 238 m currently bypass Townsville to pick up and drop off freight for Northern Australia at southern ports, which is then transported north by road).

3.7.1.2 Rail

As in Europe and North America, rail provided Australia's first terrestrial mass transport opportunity, and establishment of a 'rail head' was an indication that a settlement had 'made it'. While operated initially by private companies, a shortage of speculation capital resulted in continued rail development being undertaken by colonial governments to connect the hinterland with the major export seaports which, in most cases, were the capital cities. Planners gave little thought to connecting their railways with the other rail systems. By federation in 1901, all states except Western Australia were linked by rail, and more than 20,000 km of track had been laid. Sadly, those who envisaged a nation had not contemplated a national rail network, with three different rail gauges in use (Dept of Infrastructure & Regional Development 2014).

The construction of a transcontinental rail link between Adelaide and Darwin began in 1878, following the route taken by John McDouall Stuart during his 1862 crossing of Australia. But it was not till 1929 that the rail extension reached Alice Springs and replaced the Afghan camel train service (leading to the line's nickname *the Ghan*), by which time the line was running at a financial loss and plans for connection to Darwin were put on indefinite hold (AustralAsia Railway Corporation 2014). A 1999 economic analysis estimated extension of the rail line to Darwin would add $600 million to Australia's GDP during construction and $4.5 billion in net present value over 25 years (Bannister 2000, p. 88), and line construction began again in July 2001, with the first passenger train finally reaching Darwin on 4 February 2004 – 126 years in total and at a cost of $1.3 billion. The project was believed to be the second largest civil engineering project in Australia and the largest in the 50 years since the creation of the Snowy Mountains Scheme (built 1949–1974). In addition to heralding a new era of tourism in the Northern Territory and providing better access to and for Aboriginal communities in the region, justification for the rail link was that it would enable Darwin to serve as another trade link with Asia.

In 1991 Australia's Industry Commission (now the Productivity Commission) examined the extent to which Australia's railways acted as a monopoly and what affect this had on productivity and efficiency, and whether targeted governance reforms could produce better results. The reform of Australia's rail monopolies was estimated to unlock $5 billion per annum in productivity dividends to the Australian community. Today there is a clear organisational delineation between social and economic functions of the railways, but both functions are pursued through commercial principles and contractual arrangements. This enables social train services to be provided on tracks which have an overwhelmingly commercial purpose and commercial train services to be provided on tracks which the community perceives to be there for social reasons (Fraser 2012).

However, there is still no interstate connectivity for Queensland rail services, and when coal is excluded, rail provides just 9% indicative freight movement in Queensland compared to 88% by road (Keyte 2014). There is provision in the Developing Northern Australia White Paper for investigation of a Mt. Isa to Tennant Creek rail line, which theoretically would enable circular rail connectivity from Sydney, Adelaide, Darwin and Townsville and back down the east coast. However, there is concurrently a proposal for a dedicated freight rail corridor from Toowoomba (in South East Queensland) to the Port of Brisbane which has achieved support from all three levels of government and is likely to divert rail freight focus away from the north.

3.7.1.3 Road

In the 7 years from 2001 to 2008, Australian agriculture produced $253.5 billion in gross value agricultural commodities, almost all of which began their journey to domestic and export markets on Australia's rural roads (Fraser 2010). There are far more roads than there ever were railways in Australia – over 800,000 km valued at somewhere over $100 billion – but Australia has never addressed the question of the road system's natural monopoly in the same manner as the Industry Commission enquiry into rail monopolies (Fraser 2012). A 2012 investigation into economic reform of Australia's road sector found that:

> Road use is not directly charged. This means there is no monetary estimation of loss or gain, and a strong ability to cross-subsidise. It also means that road use, and potentially provision, is excessive at least in some places. Evidence of this includes congestion. The extent to which transport is induced by road provision is unclear. Among the issues this generates is externalities and induced car traffic. In other places it is clear the condition of roads is inadequate, and this includes local roads. Recent estimates suggest a life cycle funding gap on local roads alone of between 2 and 3 billion dollars annually. (Fraser 2012, p. 12)

In contrast to this predominately state-based system of road development, Commonwealth participation in 'beef road' development began in 1949 – a programme of improvements to facilitate cattle transport in Queensland, Western Australia and the Northern Territory with the objective of increasing beef exports as part of the Fifteen Year Meat Agreement with the UK (Department of National Development 1968). The major part of the expenditure was provided by the Commonwealth as grants – $4.332 million between 1949 and 1954. A new programme of development commenced in 1961 following consideration by the Commonwealth Government of major developmental projects which might serve to increase Australia's export income as well as provide an opportunity for the Commonwealth to be more closely associated with productive developmental projects in the outlying areas of Australia. A grant of $10 million to Queensland in 1961 was augmented by a $6.6 million loan in 1962; Western Australia received $6.9 million over a 5-year period commencing 1961; and the Northern Territory received $9.14 million for specified roads. Following approaches by state governments for additional funds in 1964, the Commonwealth directed that a comprehensive benefit-cost analysis of beef road development be undertaken. However, there was a specific statement in the 1965–1966 Budget Speech to the effect that no cut-off in the beef road programme was intended, and in November 1967 $50 million over 7 years in the form of nonrepayable grants was announced and that the grants would not be conditional on matching state contributions, although it was expected that states would continue to make additional contribution to the development of beef roads.

In May 2014, the Queensland Government committed to progress alternative inland routes as part of their addressing 'down times' on the Bruce Highway because of natural disasters. Key roads identified in these initiatives which are earmarked for potential upgrade include the Kennedy and Gregory Development Roads (RDA 2014), all originally 'beef roads'. A $260.5 million Australian and Queensland Government package providing improved Cape York access for freight, tourists and other road users will be completed June 2019, with benefits including improved safety, reduced ongoing road maintenance costs, improved community infrastructure and employment, training and business development opportunities for Indigenous and non-Indigenous people.

3.7.1.4 Air

The Air Freight Services industry transports time-sensitive and valuable goods including perishable foods, within Australia and internationally. While only 1% of exports by volume go in aircraft, because they tend to be the most expensive goods, they account for 35% of global trade by value. Although many major airlines maintain a dedicated fleet of freight aircraft, most airfreight is transported in passenger aircraft's cargo holds. The growth in international and domestic passenger flights has boosted freight capacity, putting downward pressure on freight rates. However, strong and rising outbound freight volumes have offset lower freight rates. Increased outbound freight volumes have been driven by a depreciating Australian dollar, supporting demand for a range of Australian exports, such as agricultural produce.

International food supply chains using airfreight are increasing in importance, and the value of Australia's airfreight food exports was $1.6 billion in 2011–2012 (Nguyen et al. 2013). While a cost-effective option for high-value, low-volume food where quality (and price) is dependent on timely delivery, access to airfreight is limited in northern Australia. Future growth will be influenced by market access and biosecurity policy arrangement in addition to infrastructure investment. Northern Australia's proximity to Asia won't provide agricultural export benefit without direct connectivity: while 90% of airfreight is carried in the bellies of passenger jets, presently only 1.5% of international passengers arrive/depart directly from Northern Australia (Sprigg 2015).

3.7.2 Communication Infrastructure

Communication services within Australia post-European settlement were the almost exclusive responsibility of government. The first regular postal service began in 1821 in New South Wales; other colonies followed and soon Australia's mail network was entirely run and regulated by government. This began a pattern that was to dominate the communication industry for the next 150 years – almost complete government control of Australia's communications services (Australian Bureau of Statistics Yearbook 2001).

When the telegraph first appeared in Europe in 1844, Australia was quick to adopt the new technology, and by the mid-1860s all regional centres in the south-east of the country were part of a virtually instantaneous communications network owned, maintained and managed by the government. The final and most significant breakthrough came in 1872, when Stuart's crossing of the Northern Territory enabled the establishment of Australia's first international telecommunications system – a telegraph link to Asia. This in turn connected Australia to the European and American lines, ending its isolation from the rest of the world.

Telephones quickly followed the telegraph, and in 1882 the Sydney telephone exchange made personal communication available to the average Australian, just 6 years after Alexander Graham Bell took out his patent. Ironically, it was often easier for a Sydney caller to reach London than outback New South Wales, so in 1960 the Postmaster General (PMG) made a firm policy commitment: while Australia's international telecommunications industry would continue to be developed, the focus would be upon providing modern communication services to all Australians.

Television broadcast from Sydney began in 1956, and 6 years later television was available in all capital cities except Darwin. The PMG experimented with data services, sending computerised stock exchange and business information over the telephone system in 1964. The Overseas Telecommunications Commission (OTC) was established in 1946 to oversee Australia's international telecommunications services and their development. When the first international communications satellite, INTELSAT I, linked North America and Europe in 1965, it became clear that ground-based technology was no longer enough to connect Australia to the world. In 1966 INTELSAT II was launched, providing a satellite link between Australia and the international telecommunications network, and by 1968 the entire Australian telecommunications system was 'plugged' into this network. By 1987 all areas in Australia had basic telephone services, no matter how remote. Australia had achieved telecommunications maturity, with all Australians linked by a single (government-owned) infrastructure called Telecom (later Telstra).

Worldwide, information and telecommunications industries changed rapidly in the next decade, and the quality of Australia's domestic telecommunication services was crucial for the nation's participation in this new Internet economy. However, the required technology developments of the Internet age brought concerns that Australia would be unable to compete unless it changed its communications industry structure to achieve market growth, and competition was seen as the remedy. This required sweeping changes to the established industry that carried significant risk – if competition was introduced too quickly and without due care, new entrants would flounder, causing the entire Australian telecommunications industry to become destabilised. In 1997 there was a partial sale of Telstra, and the industry opened to full competition, and all limitations on the number of licenced players were removed, and anticompetitive conduct was prohibited. In 1999, further sales brought the privately owned portion of Telstra to 49.9%.

In 2010, NBN Co (National Broadband Network Company, again a government-owned corporation wholly owned by the Commonwealth of Australia as a government business enterprise) was established to design, build and operate a National Broadband Network to deliver a high-speed Internet throughout Australia. As Tickell (2011, p. 931) describes:

> [Australia is] now living in a more globalized world of rapid communication. There is already a kind of universal language of electronics. Ideas, units of information—or memes—will pass almost instantaneously between countries, communities and individuals. The wiring of the planet with fibre optics, cellular wireless, satellites and digital television is transforming human relationships. For the first time, there will be something like a single human civilization. More than ever humans can be regarded, like certain species of ants, as a super-organism'

The enhanced communication opportunities, access to information and social networking that the Internet provides have changed many aspects of Australian society, particularly in ways that circumvent physical social isolation. This is particularly important in rural and remote areas, where the Internet can help ameliorate the tyranny of distance by providing social connections and also provide mechanisms to access knowledge about health and wellbeing (Scott et al. 2015). This revolution in information technologies is making it possible for people to collaborate globally in real time and enhancing the capacity to respond to some of the dangers associated with other forms of globalisation (Friedman 2005).

However, the *Regional Telecommunications Review 2015* (RTIRC) found that while regional Australians have a higher dependency on mobiles than their urban counterparts because of the broader geographic range within which many conduct their working and everyday lives, the quality and extent of mobile coverage continues to be a major concern. Even though Australians enjoy one of the highest penetrations of mobile broadband in the world, with smartphone penetration in Australia expected to surpass 90% by the end of 2018, while the rest of the world will take until 2023 (Economics 2018)

> the low population density … means that new approaches are needed to assess the priorities of those in the 70% of Australia's land mass that has no mobile coverage, and to improve poor coverage elsewhere. (RTIRC 2015, p. ix)

3.7.3 Water Infrastructure

As residents of the driest continent on Earth, securing a reliable water supply should be at the forefront of planning considerations for securing Australia's future prosperity. There are more than 820 dams in Australia, with a total capacity exceeding 91,000 gl[16] (ANCOLD 2012). However, most dams are at southern latitudes, where Mediterranean climates mean water storage is required to last through dry summers. Less than 10% of Australia's dams are in Northern Australia.

In the lead up to a national election, Australia's then Prime Minister Kevin Rudd announced in August 2013 his plan for Northern Australia and called for a 'national imagination' to take advantage of the 'enormous agricultural potential' of the Top End, including harnessing 'the bountiful supply of water', while his political opponent (and subsequent Prime Minister) Tony Abbott had already committed to investigating establishment of a Water Project Development Fund 'to support the

[16] A gigalitre = 1000 ml or 1000,000,000 l.

advancement of meritorious proposals for water infrastructure across Northern Australia, including dams and groundwater projects' (Liberal Party of Australia 2013). However Rayner (2013, p. 2) warned that the Prime Minister's announcements echoed the:

> European-derived understanding of rivers that was used when settling the Murray-Darling Basin. There, early decisions to allocate water for agricultural progress were made implicitly – water licences were handed out and the system quickly became over-allocated. As consumptive capacity was taken up, ecological decline accelerated and "fitful, reluctant co-operation" in water management began.

Rayner's view was in alignment with the 2009 report by the Northern Australia Land and Water Science Review (NALW) (CSIRO 2009) that, despite Northern Australia's rainfall, there may be only 600 gl of water available across Northern Australia that could support new consumptive use, a volume adequate to irrigate only 40,000–60,000 ha of intensive agriculture. This finding was very much at odds with both earlier research and contemporary industry expectations: Gifford et al. (1975) had estimated the potential irrigable area in Northern Australia to be twice that of Southern Australia and that only 5% of this area had been developed compared with 89% of potential southern areas. While Nothrup (1981) found this to be an overestimation, his revised figure was 13.3 million ha of potential agricultural land. The Integrated Food and Energy Developments proposal for the Gilbert River alone was for 50,000 ha of irrigated cropping land (IFED 2014).

A subsequent Australian Government commissioned 2.5 year/$15 million CSIRO Northern Australia Water Resource Assessment identified that:

- In the Fitzroy catchment of Western Australia, water harvesting (water pumped into ring tanks) could potentially support 160,000 ha growing one dry season crop a year in 85% of years. Independent of surface water, groundwater could potentially support up to 30,000 ha of hay production in all years.
- In the Darwin catchments, a combination of major dams, farm-scale off-stream storage and groundwater could potentially support up to 90,000 ha of dry season horticulture and mango trees.
- In the Mitchell catchment of Queensland, large instream dams could potentially support 140,000 ha of year-round irrigation. Alternatively, water harvesting could potentially enable up to 200,000 ha, growing one dry season crop per year.
- If irrigated opportunities were pursued to their fullest extent, they would only occupy about 3% of the assessment area.

(It is available at https://www.csiro.au/en/research/major-initiatives/northern-australia/current-work/nawra/overview, and an interactive explorer enabling land suitability, soil, water, climate, ecology and Indigenous interests is available at https://nawra-explorer.csiro.au/.)

The discrepancy between these and the earlier estimation was explained by the previous study only covering 155,000 km^2 (approximately 5%) of Northern Australia, a detail not explicit at the time. The debate around new dams in the north continues, with local proponents often at odds with southern experts. In an opinion piece published in *The Guardian* (2018) after release of the above NAWRA report, a senior research fellow from the University of Queensland observed that the only obstacle to a massive expansion in irrigated agriculture was economic reality, as 'agricultural economists recognised long ago that the environment in northern Australia was not good for irrigated agriculture. The converse recognition, that irrigation schemes are often disastrous for the environment, came much later'. The important point to note here is that projections are possibly more influenced by the aspirations of the commissioning proponent than the science on which they base their estimates.

As it stands, the only schemes built expressly for irrigation in Northern Australia are:

1. The 17,000 ha Mareeba-Dimbulah Irrigation Scheme on Queensland's Atherton Tablelands – serviced by Tinaroo Dam, completed in 1958.
2. The Ord River Irrigation Scheme in Western Australia – serviced by Lake Argyle, Australia's largest artificial lake (by volume) and retained by the Ord River Dam built in 1971 by the American Dravo Corporation.
3. The 103,000 ha Burdekin Irrigation Scheme in Queensland – serviced by Lake Dalrymple and retained by the Burdekin Falls Dam, Stage 1 of which was completed by Leighton Holdings in 1987. The design allows for future storage capacity increases, and the dam is also capable of providing water for Townsville City. This was utilised during 2017 when, after three failed wet seasons, the city's Ross Dam water supply fell below 15%. The contemporaneous public water security debate reinvigorated consideration of the Burdekin Falls Dam additional stages and other water storage options.

The way water is used in Australia is changing: along with improved understanding of the value of water has come pressure for better management of this limited resource, and past management strategies may no longer be enough. Along with increased mining activity, outback areas are diversifying their traditional economic bases into service industries including tourism and irrigated agriculture, and Traditional Owners are asserting their water use and access rights (Larson 2006). These changes have implications for sustainable water management, which must cater for the individuals and groups who rely upon water for their life and livelihoods.

3.7.4 Infrastructure Summary

Many of Bauer's (1977) identified impediments to agricultural success have been addressed and resulted in improved quality of life and service provision for northern residents. However, the provision of infrastructure continues to be a significant challenge and topic of public debate, particularly where a sparse population relies on public provision of transport, energy, water and communications to an even greater extent than those in the more populated south. While the Northern Australian economy has grown faster than the rest of Australia in recent years, and its demographic profile is also younger and faster growing, the cost of infrastructure provision remains high. Although cost-benefit analysis is an essential consideration of public investment prioritisation, it should never be the sole consideration as its 'focus on expected or known infrastructure demand can be inadequate where future opportunities are central to the business case' (Infrastructure Australia 2015, p. 27).

Improved infrastructure attracts people, who demand improved infrastructure. This chicken-and-egg relationship is not unique to Northern Australia, nor is it easily rectified. Federal, state and territory governments aim to promote economic growth and in some cases population growth, through policy settings and regional plans. Accordingly, the 2015 *Northern Australia Audit* (Infrastructure Australia 2015) assesses critical infrastructure gaps under both policy- and plan-driven economic and population growth scenarios. However, the audit focuses primarily on infrastructure relating to larger population centres (more than 3000 persons), as well as to areas of significant existing or prospective economic activity. In consequence, the essential infrastructure needs of the many smaller communities, and in particular remote Indigenous communities, fall outside the audit's scope.

3.8 Are We There Yet?

While infrastructure will continue to be an essential part of the agriculture debate in Northern Australia, history demonstrates that it is neither an absolute limitation to new agricultural development nor a panacea for addressing slow rates of development. Thinking about this raises new questions: What is a slow rate of development? By whose standards should *fast* and *slow* be measured? Do *fast* and *slow* necessarily equate to *good* and *bad* or even to *acceptable* and *unacceptable*?

European settlers arrived in Australia during a period of immense social change coupled with unprecedented industrial innovation in Europe, and around the world this rate of change has not abated. Australia was seen as a land of big horizons, and bigger possibilities, and it was well placed to capitalise on the benefits of mechanised agriculture. The most commonly used international measure of a country's performance continues to be its gross domestic product,[17] and since agriculture remains an important contributor to Australia's GDP, it is understandable that a popular attitude of fast development being desirable continues. All recent Australian government elections have seen growth touted by major political parties as a desirable goal, and it is a sentiment readily supported and promulgated by popular news media.

But the world is changing. As the population grows, humanity is becoming increasingly aware of the limit to Earth's resources and the need for sustainable management. In terms of human land use, Australia is both a young country and an old country, and it is just possible that this paradox contributes to the confusion over agricultural expansion in Northern Australia. The next chapter will consider some of these influencers.

[17] The OECD defines GDP as an aggregate measure of production equal to the sum of the gross values added of all resident and institutional units engaged in production (plus any taxes, and minus any subsidies, on products not included in the value of their outputs) (OECD 2002).

References

AACo. (2012). AACo Board approves construction of Northern Territory meat processing facility. Retrieved from www.aaco.com.au/investors-and-media/media/aaco-board-approves-construction-of-northern-territory-meat-processing-facility/.

ABARES. (2015). *Agricultural commodities: September quarter 2015*. Retrieved from Canberra: www.agriculture.gov.au/abares/publications.

ABS. (2014). *Agricultural land and water ownership, June 2013* (7127.07). Retrieved from Canberra: http://www.abs.gov.au/AUSSTATS/abs@.nsf/Lookup/7127.0Explanatory%20Notes1Jun%202013?OpenDocument.

ABS. (2015). *Regional population growth, Australia, 2013–14*. Retrieved from Canberra: www.abs.gov.au/ausstats/abs@.nsf/mf/3218.0.

Allen, P. (2013). Facing food security. *Journal of Rural Studies, 29*(0), 135–138. https://doi.org/10.1016/j.jrurstud.2012.12.002.

Ambler-Edwards, S., Bailey, K., Kiff, A., Lang, T., Lee, R., Marsden, T., … Tibbs, H. (2009). *Food futures: Rethinking UK strategy*. Retrieved from Chatham House, London. http://www.chathamhouse.org/sites/default/files/public/Research/Global%20Trends/r0109foodfutures.pdf.

ANCOLD. (2012). Register of large dams in Australia. Retrieved from www.ancold.org.au/?page_id=24.

Aschmann, H. (1977). Views and concerns relating to northern development. *North Australia Research Bulletin, 1*, 31–57.

AustralAsia Railway Corporation. (2014). History of the railway. Retrieved from www.aarail.com.au/

Australian Bureau of Statistics. (2001). *1301.0 – year book Australia*. Retrieved from www.abs.gov.au/ausstats/abs@.nsf/Previousproducts/1301.0Feature%20Article432001.

Australian Food News. (2011). No relaxation on quarantine for imported bananas. From www.ausfoodnews.com.au/2011/02/08/no-relaxation-on-quarantine-for-imported-bananas.html.

Australian Government. (2012). *Australia in the Asian century: White paper*. Canberra: Department of the Prime Minister and Cabinet. Retrieved from www.defence.gov.au/whitepaper/2013/docs/australia_in_the_asian_century_white_paper.pdf.

Australian Government. (2015a). *Agricultural competitiveness white paper.* Canberra Retrieved from http://agwhitepaper.agriculture.gov.au/SiteCollectionDocuments/ag-competitiveness-white-paper.pdf.

Australian Government. (2015b). *Our north, our future: White paper on developing Northern Australia*. Canberra: Commonwealth of Australia. Retrieved from http://northernaustralia.infrastructure.gov.au/white-paper/files/northern_australia_white_paper.pdf.

Ball, G. (2012). Whole of value chain scenario for Asia-Pacific food leadership. *Farm Policy Journal, 9*(4), 1–9.

Bannister, L. (2000). 'On track': The AustralAsia railway. In R. Dixon (Ed.), *Business as usual? Local conflicts and global challenges in northern Australia* (pp. 87–91). Darwin: North Australia Research Unit.

Barr, N. (2009). *The house on the hill: The transformation of Australia's farming communities*. Canberra: Land & Water Australia.

Bauer, F. H. (1977). *Cropping in North Australia: Anatomy of success and failure*. Paper presented at the 1st North Australia Research Unit Seminar., Darwin N.T.

Bauer, F. H. (1985). A brief history of agriculture in north-west Australia. In R. C. Muchow (Ed.), *Agroresearch for the semi-arid tropics: North-west Australia* (pp. 12–31). St. Lucia: Uni. of Qld Press.

Bendle, M. F. (2013). Remaking Australia as a frontier society. *Quadrant, 57*(5), 7–13.

Bennett, M. M. (1927). *Christison of Lammermoor*. London: Alston Rivers Ltd.

Bindraban, P. S., & Rabbinge, R. (2012). Megatrends in agriculture – views for discontinuities in past and future developments. *Global Food Security, 1*(2), 99–105. https://doi.org/10.1016/j.gfs.2012.11.003.

BITRE. (2009). *Northern Australia statistical compendium 2009*. Canberra: Bureau of Infrastructure, Transport and Regional Economics.
Blight, D. (2012). *Overview*. Paper presented at the Crawford Fund 2012 Parliamentary Conference.
Bottoms, T. (2013). *The conspiracy of silence: Queensland's frontier killing times*. Retrieved from http://jcu.eblib.com.au/patron/FullRecord.aspx?p=1190541.
Bowler, I. (1979). *Government and agriculture: A spatial perspective*. London: Longman Group Ltd.
Burch, D., Goss, J., Lawrence, G., & Rickson, R. E. (1999). Introduction. The global restructuring of food and agriculture: Contingencies and parallels in Australia and New Zealand. *Rural Sociology, 64*(2), 179–185. https://doi.org/10.1111/j.1549-0831.1999.tb00012.x.
Carment, D. (1996). *Looking at Darwin's past: Material evidence of European settlement in tropical Australia*. Darwin: North Australia Research Unit. The Australian National University.
CEDA. (2013). International success as important as domestic factors for Aust. economy in Asian Century. Retrieved from www.ceda.com.au/.
COAG. (2011). *National ports strategy*. Retrieved from www.infrastructureaustralia.gov.au/publications/files/COAG_National_Ports_Strategy.pdf.
Cocco, R. (2013). [Pers. Comm. re. Project Catalyst].
Collits, P. (2006). Great expectations: What regional policy can achieve in Australia. *Farm Policy Journal, 3*(3), 1–11.
Collits, P. (2011). Country towns in a big Australia: The decentralisation debate revised. In J. Martin & T. Budge (Eds.), *The sustainability of Australia's country towns: Renewal, renaissance, resilience* (pp. 23–57). Ballarat: VURRN Press.
Cook, G. D. (2009). Historical perspectives on land use development in northern Australia: With emphasis on the Northern Territory. In *Northern Australia land and water science review full report October 2009* (pp. 30). Department of Infrastructure, Transport, Regional Development and Local Government.
Cook, G. D., Jackson, S., & Williams, R. J. (2012). A revolution in northern Australian fire management: Recognition of indigenous knowledge, practice and management. In R. A. Bradstock & R. J. Williams (Eds.), *Flammable Australia* (pp. 293–305). Collingwood: CSIRO Publishing.
Crothers, A. (2015). Stanbroke stitches up cotton plans. *Queensland Country Life*.
Cruz, M. (2010). A living space: The relationship between land and property in the community. *Political Geography, 29*(8), 420–421. https://doi.org/10.1016/j.polgeo.2010.09.003.
CSIRO. (2009). *Northern Australia land and water science review 2009 full report*. Retrieved from http://nalwt.gov.au/files/00_Executive_Summary_and_Introduction.pdf.
CSIRO. (2013). *Land tenure in northern Australia: Opportunities and challenges for investment*. Retrieved from http://regional.gov.au/regional/ona/land-tenure/pdfs/land-tenure-20130717.pdf.
DAFF. (2013). *National Food Plan: Our food future*. Canberra: Department of Agriculture, Fisheries and Forestry. Retrieved from daff.gov.au/nationalfoodplan.
Deloitte Access Economics. (2016). *Building resilient infrastructure*. Retrieved from http://australianbusinessroundtable.com.au/assets/documents/Report%20-%20Building%20Resilient%20Infrastructure/Report%20-%20Building%20resilient%20infrastructure.pdf.
Department of National Development. (1968). *Beef roads scheme*.
Dept of Infrastructure & Regional Development. (2014). History of rail in Australia. Retrieved from www.infrastructure.gov.au/rail/trains/history.aspx.
Durack, M. (1968). *Kings in grass castles*. (Repr. with new introd ed. London: Constable.
Economics, D. A. (2018). *Technology, media and telecommunications predictions*. Retrieved from https://www2.deloitte.com/cn/en/pages/technology-media-and-telecommunications/articles/tmt-predictions-2018.html.
Edwards, G. (2018). *From red to green to black: A stewardship incentives scheme for improving Queensland's pastoral lands*. A submission to The Royal Society of Queensland.
Ericksen, P. J., Ingram, J. S. I., & Liverman, D. M. (2009). Food security and global environmental change: Emerging challenges. *Environmental Science & Policy, 12*(4), 373–377. https://doi.org/10.1016/j.envsci.2009.04.007.

Fisher, M. J., Garside, A. L., Skerman, P. J., Chapman, A. L., Strickland, R. W., Myers, R. J. K., … Henzell, E. F. (1977). *The role of technical and related problems in the failure of some agricultural development schemes in Northern Australia.* Paper presented at the 1st NARU Seminar, Darwin N.T.

Food & Agriculture Organisation. (2015). *The state of food insecurity in the world 2015.* Retrieved from Rome: http://www.fao.org/3/a4ef2d16-70a7-460a-a9ac-2a65a533269a/i4646e.pdf.

Fraser, A. (2008). Gaining ground: Emerging agrarian political geographies. *Political Geography, 27*(7), 717–720. https://doi.org/10.1016/j.polgeo.2008.08.007.

Fraser, L. (2010). *Going nowhere: The rural local road crisis. Its national significance and proposed reforms.* Retrieved from http://www.infrastructureaustralia.gov.au/publications/files/Australian_Rural_Roads_Group_Report.pdf.

Fraser, L. (2012). *Economic reform of Australia's road sector: Precedents, principles, case studies and structures.* Retrieved from http://infrastructureaustralia.gov.au/policy-publications/publications/files/Competition_Reform_of_the_Road_Sector.pdf.

Friedman, T. L. (2005). *The world is flat: A brief history of the twenty-first century.* New York: Farrar, Straus and Giroux.

Gammage, B. (2011). *The biggest estate on earth: How aborigines made Australia.* Crows Nest: Allen & Unwin.

Garnett, S., Woinarski, J., Gerritsen, R., & Duff, G. (2008). *Future options for North Australia.* Darwin: Charles Darwin University Press.

Gifford, R. M., Kalma, J. D., Aston, A. B., & Millington, R. (1975). Biophysical constraints in Australian food production: Implications for population policy. *Search, 6,* 212–223.

Gilmore, M. A. (2005). *KILL, CURE OR STRANGLE: The history of government intervention in three agricultural industries on the Atherton Tablelands, 1895–2005.* (Doctor of Philosophy), James Cook University.

Gray, G. (2009). The importance of northern Australia. *Farm Policy Journal, 6*(2), 1–9.

Gray, E., Oss-Emer, M., & Sheng, Y. (2014). *Australian agricultural productivity growth: Past reforms and future opportunities.* Canberra: Australian Bureau of Agricultural and Resource Economics and Sciences. Retrieved from http://data.daff.gov.au/data/warehouse/9aap/2014/apgpfd9abp_20140220/AgProdGrthPstRfmFtrOppsv1.0.0.pdf.

Habermas, J. (2010). In B. Fultner (Ed.), *Jurgen Habermas: Key concepts.* Durham: Acumen.

Hajkowicz, S., & Eady, S. (2015). *Megatrends impacting Australian agriculture over the coming twenty years.* Retrieved from [online].

Hathaway, D. E. (1969). The search for new international arrangements to deal with the agricultural problems of industrialised countries. In V. Papi & C. Nunn (Eds.), *Economic problems of agriculture in industrial societies* (pp. 51–69). London: Macmillan.

Heathcote, R. L. (1975). *The world's landscapes: Australia.* London: Longman.

Heathcote, R. L. (1994). *Australia* (2nd ed.). England: Longman Scientific & Technical.

Hoath, A., & Pavez, L. A. (2013). *Survey report: Intersections of mining and agriculture, Boddington Radius: Land use, workforce & expenditure patterns.* Retrieved from http://apo.org.au/sites/default/files/docs/CSIRO_mining_and_agriculture_2013.pdf.

Hogan, A., & Young, M. (2013). Visioning a future for rural and regional Australia. *Cambridge Journal of Regions, Economy and Society, 6*(2), 319–330. https://doi.org/10.1093/cjres/rst005.

Hughes, M. (2014). The role of a national food brand for Australia. *Farm Policy Journal, 11*(4), 11–19.

Hull, C. (2000). Governance and the economic story of the north. In R. Dixon (Ed.), *Business as usual? Local conflicts and global challenges in Northern Australia* (pp. 43–69). Darwin: North Australia Research Unit.

IFED. (2014). Etheridge integrated agricultural project. Retrieved from http://i-fed.com.au/project/.

Infrastructure Australia. (2015). *Northern Australia Audit: Infrastructure for a developing north.* Retrieved from Canberra: http://infrastructureaustralia.gov.au/policy-publications/publications/files/IA_Northern_Australia_Audit.pdf.

JCU. (2014). *State of the tropics: 2014 report*. Retrieved from Townsville: www.stateofthetropics.org.

Kaplan, H. B. (2006). Old states, new threats. Retrieved from www.militaryphotos.net/forums/showthread.php?79192-Old-States-New-Threats-RD-Kaplan-article.

Kaplan, H. B. (2009). The revenge of geography. *Foreign Policy, 172*, 96–105.

Keogh, M. (2009). Agriculture in northern Australia in the context of global food security challenges. *Farm Policy Journal, 6*(2), 35–43.

Keogh, M. (2012a). Editorial. *Farm Policy Journal, 9*(4), iv.

Keogh, M. (2012b). *Further evidence that Australian financial markets just don't 'get' agriculture*. Retrieved from http://www.farminstitute.org.au/ag-forum/further_evidence_that_australian_financial_markets_just_dont_get_agriculture.

Keogh, M. (2014). Optimising Australian agriculture's comparative advantage. *Farm Policy Journal, 11*(3), 1–7.

Keyte, P. (2014). *Inland rail – getting started*. Paper presented at the Qld Agriculture Conference, Brisbane.

Lagi, M., Bertrand, K. Z., Bar-Yam, Y. (2011). The food crises and political instability in North Africa and the Middle East. *Social Science Research Network*.

Larson, S. (2006). *Analysis of the water planning process in the Georgina and Diamantina catchments: An application of the institutional analysis and development (IAD) framework*. Available from http://citeseerx.ist.psu.edu/viewdoc/download?doi=10.1.1.533.2896&rep=rep1&type=pdf.

Lawrence, G., Richards, C. A., & Cheshire, L. (2004). The environmental enigma: Why do producers professing stewardship continue to practice poor natural resource management? *Journal of Environmental Policy & Planning, 6*(3–4), 251–270. https://doi.org/10.1080/1523908042000344069.

Lawrence, G., Richards, C., & Lyons, K. (2013). Food security in Australia in an era of neoliberalism, productivism and climate change. *Journal of Rural Studies, 29*(0), 30–39. https://doi.org/10.1016/j.jrurstud.2011.12.005.

Liberal Party of Australia. (2013). Real solutions for all Australians: The direction, values and policy priorities of the next Coalition Government. [online] www.realsolutions.org.au.

Maher, S. (2011, 17 September). Tony Abbott's plan for northern foodbowl. *The Australian*.

Marginson, S. (1997). *Markets in education*. St. Leonards: Allen and Unwin.

Martin, J., Budge, T., Andrew, B. (2011). Introduction. In *The sustainability of Australia's country towns: Renewal, renaissance and resilience* (pp. 1–9). Ballarat, Australia: URRN Press.

Maye, D., & Kirwan, J. (2013). Food security: A fractured consensus. *Journal of Rural Studies, 29*(0), 1–6. https://doi.org/10.1016/j.jrurstud.2012.12.001.

McCrone, R. G. L. (1962). *The economics of subsidising agriculture*. London: George Allen and Unwin.

McGregor, R. (2016). *Environment, race, and nationhood in Australia: Revisiting the empty north*. New York: Palgrave Macmillan.

McLean, D., & Gray, I. (2012). History for policy: What is the value of history for the in-principle assessment of government intervention in rural Australia? *Rural Society, 21*(3), 190–197.

Misselhorn, A., Aggarwal, P., Ericksen, P., Gregory, P., Horn-Phathanothai, L., Ingram, J., & Wiebe, K. (2012). A vision for attaining food security. *Current Opinion in Environmental Sustainability, 4*(1), 7–17. https://doi.org/10.1016/j.cosust.2012.01.008.

Mollah, W. S. (1980). North Australian cropping studies, III. The tipperary story: An attempt at large-scale grain sorghum development in the Northern Territory. *North Australia Research Bulletin, 7*, 59–191.

Mooney, P. H., & Hunt, S. A. (2009). Food security: The elaboration of contested claims to a consensus frame*. *Rural Sociology, 74*(4), 469–497. https://doi.org/10.1111/j.1549-0831.2009.tb00701.x.

Moritz, C., Ens, E., Potter, S., & Catullo, R. (2013). The Australian monsoonal tropics: An opportunity to protect unique biodiversity and secure benefits for Aboriginal communities. *Pacific Conservation Biology, 19*(3/4), 343–355.

Müller, C., & Lotze-Campen, H. (2012). Integrating the complexity of global change pressures on land and water. *Global Food Security, 1*(2), 88–93. https://doi.org/10.1016/j.gfs.2012.11.001.
Neutze, G. M. (1965). *Economic policy and the size of cities*. Canberra: Australian National University.
NFF. (2012). *National farmers' federation Blueprint for Australian agriculture 2013–2020*. Retrieved from www.nff.org.au/blueprint.
Ng, M., Fleming, T., Robinson, M., Thomson, B., Graetz, N., Margono, C., … Gakidou, E. (2014). Global, regional, and national prevalence of overweight and obesity in children and adults during 1980–2013: A systematic analysis for the global burden of disease study 2013. *The Lancet, 384*(9945), 766–781. doi:https://doi.org/10.1016/S0140-6736(14)60460-8.
Nguyen, N., Hogan, L., Lawson, K., Gooday, P., Green, R., Harris-Adams, K., & Mallawaarachchi, T. (2013). *Infrastructure and Australia's food industry: Preliminary economic assessment*. Canberra: Australian Government.
Nothrup, L. (1981). Potential: Water resources. In Commonwealth Council for Rural Research and Extension (Ed.), *Rural research in Northern Australia*. Canberra: Australian Government Publishing Service.
O'Faircheallaigh, C. (2011). Green black conflict over gas development in the Kimberley: A sign of things to come? Retrieved 31 Aug., 2014 https://theconversation.com/green-black-conflict-over-gas-development-in-the-kimberley-a-sign-of-things-to-come-3539.
O'Meagher, B. (2005). Policy for agricultural drought in Australia: An economics perspective. In L. Botterill & D. Wilhite (Eds.), *From disaster response to risk management* (Vol. 22, pp. 139–155). Netherlands: Springer.
OECD. (2002). Glossary of statistical terms. Retrieved from http://esa.un.org/unsd/sna1993/introduction.asp.
OECD. (2015). The organisation for economic co-operation and development: Better policies for better lives. Retrieved from www.oecd.org/about/.
Owens, J. (2013, July 16, 2013). Northern foodbowl needs gas. *The Australian, July 16*.
Pieck, S. K., & Moog, S. A. (2009). Competing entanglements in the struggle to save the Amazon: The shifting terrain of transnational civil society. *Political Geography, 28*(7), 416–425. https://doi.org/10.1016/j.polgeo.2009.09.009.
Port Master Planning. (2014). Ports Australia. Retrieved from www.portsaustralia.com.au/tradestats/.
Power, N. (2013, 5–6 January). Mine jobs end Aboriginal hardship. *The Weekend Australian*.
Prasad, S., & Langridge, P. (2012). *Australia's role in global food security*. Canberra: Office of the Chief Scientist.
Queensland Wild Rivers Act 2005, (2005).
Rayner, T. (2013). Dam it all? River futures in northern Australia. Retrieved from http://theconversation.com/dam-it-all-river-futures-in-northern-australia-17131.
RDA. (2011). *Regional Road map – Far North Queensland & Torres Strait Region 2011–2012*. Retrieved from http://www.rdafnqts.org.au/images/pdf/RDA%20FNQTS%20Regional%20Road%20Map%202012%20-%20FINAL%20comp.pdf.
RDA. (2014). Priority inland roads make the investment list [Press release].
RTIRC. (2015). *Regional telecommunications review 2015: Unlocking the potential in regional Australia*. Retrieved from http://www.rtirc.gov.au/wp-content/uploads/sites/2/2015/10/RTIRC-Independent-Committee-Review-2015-FINAL-Low-res-version-for-website.pdf.
Scott, J., Lyons, A., & MacPhail, C. (2015). Desire, belonging and absence in rural places. *Rural Society, 24*(3), 219–226. https://doi.org/10.1080/10371656.2015.1099263.
Self, P., & Storing, P. (1962). *The state and the farmer*. London: George Allen and Unwin.
Sen, A. (1981). *Poverty and famines: An essay on entitlement and deprivation*. Oxford: Oxford University Press.
Sorensen, A. D. (1993). The future of the country town: Strategies for local economic development. In A. D. Sorensen & W. R. Epps (Eds.), *Prospects and policies for rural Australia*. Cheshire: Longman.

Sprigg, A. (2015). *Proximity vs accessibility: Northern Australia's international connectivity.* Paper presented at the Developing Northern Australia Conference, Townsville.

Stuart, J. M. D. (1865). *Explorations in Australia: The journals of John McDouall Stuart during the years 1858, 1859, 1860, 1861 & 1862.* London: Saunders, Otley and Company.

The Guardian. (2018). Reality is the enemy of irrigated agriculture, Matt Canavan, not 'greenies' [Press release]. Retrieved from https://www.theguardian.com/commentisfree/2018/sep/12/reality-is-the-enemy-of-irrigated-agriculture-matt-cananvan-not.

Tickell, C. (2011). Societal responses to the Anthropocene. *Philosophical Transactions of the Royal Society, 369*, 926–932. https://doi.org/10.1098/rsta.2010.0302.

UNISDR. (2015). *Sendai framework for disaster risk reduction 2015–2030.* Retrieved from http://www.unisdr.org/files/43291_sendaiframeworkfordrren.pdf.

Vanclay, F. (2003). The impacts of deregulation and agricultural restructuring for rural Australia. *The Australian Journal of Social Issues, 38*(1), 81–94.

West, P. C., Gerber, J. S., Engstrom, P. M., Mueller, N. D., Brauman, K. A., Carlson, K. M., … Siebert, S. (2014). Leverage points for improving global food security and the environment. *Science, 345*(6194), 325–328. doi:https://doi.org/10.1126/science.1246067.

Western, M., Baxter, J., Pakulski, J., Tranter, B., Western, J., Egmond, M. v., … Gellecum, Y. v. (2007). Neoliberalism, inequality and politics: The changing face of Australia. *Australian Journal of Social Issues, 42*(3), 401–418.

Wikipedia: The Free Encyclopedia. (2015). Terra nullius. Retrieved from https://en.wikipedia.org/w/index.php?title=Terra_nullius&oldid=672441443.

Wilson, G. K. (1977). *Special interest and policy making.* London: John Wiley.

WWF. (2013). Project catalyst. Retrieved from www.wwf.org.au/about_us/working_with_business/project_sponsorships/project_catalyst/.

Yeatman, A. (2000). The politics of post-patrimonial governance. In T. Seddon & L. Angus (Eds.), *Beyond nostalgia: Reshaping Australian education* (pp. 170–185). Melbourne: ACER Press.

Chapter 4
Framing Northern Australian Agriculture's Future

It may be useful to think of places, not as areas on maps, but as constantly shifting articulations of social relations through time...the identity of places ... is always in that sense, temporary, uncertain and in process.
Massey (1995)

4.1 Introduction

This chapter frames Northern Australia's possible agricultural future. Land is the fundamental component of Australia's common wealth, a communal resource essential for production and life. As human populations have grown and civilisations developed, land use planning has emerged as a profession to facilitate a negotiated outcome between conflicting parties (Jones 2014). While regional land use planning has not developed to the level of complexity of urban planning, its focus in most countries

> remains on the institutional processes to meet "public" interest and "utopian" ideals, of which sustainable development is the latest manifestation ... planning not only regulates/constrains the property market, it also shapes and stimulates it. (Jones, 2014, p. 578)

Despite widespread adoption of neo-liberal principles by governments in the later decades of the twentieth century, and the commensurate improvement in living standards attributed to them, the free market has not delivered an optimal simultaneous solution for allocating resources, maximising consumer welfare, stabilising foreign trade, and reducing agricultural price instability. Governments are still called upon to intervene in the market and stimulate, regulate, or control economic forces, particularly when policy focus moves from direct production issues into less agreed arenas such as environmental management. Some of these policies emphasise the willingness of the middle-class consumer to pay a little extra for quality, a force that encourages product differentiation and thereby feeds investment in both production and marketing of new goods. This latter role has become more pronounced with the

---

Keith Noble has contributed more to this chapter.

K. Noble et al., *Agriculture and Resilience in Australia's North*,
https://doi.org/10.1007/978-981-13-8355-7_4

expansion of global trade, and new trade theories have evolved to explain why most trade expansion has been occurring at the extensive margin – that is, through the expansion of new goods rather than greater trade of existing products (Hummels and Klenow 2002).

This supports the view of Vanclay (2003) that the landscape in which agriculture occurs is a social as well as a physical landscape, and Australia's agriculture policy has included many examples of the regulation of commercial services in rural areas to ensure cost parity between rural and urban areas and government services to promote agriculture and rural development. Many rural areas though were developed only because of government policies – such as the post-world war's soldier resettlement schemes which gave land to returning soldiers, while other policies limited the area of land that could be owned and how it could be utilised. New Australian immigrants often preferentially settled in particular districts, creating an ethnic patterning. Thus the structure and overall prosperity of agriculture is socially, culturally, politically, economically and historically shaped (Vanclay 1997) by both domestic and international factors, which Farmar-Bowers (2011, p. 159) describes as the confluence of *planet-matters*[1] and *people-matters*[2].

This chapter further considers these *planet and people matters*, particularly their interplay and specific relevance to the past, present and future individuals involved in Northern Australian agriculture.

4.2 Australian Environmentalism and the Emergence of Sustainability

In 1989, 200 years after the European settlement of Australia and on the doorstep of the *Decade of Land Care*, the (then) Australian Prime Minister Bob Hawke launched a 10-year Commonwealth Government commitment to Australia's environmental protection, titled *Our Country, Our Future*. This first ever comprehensive statement on the environment by an Australian Prime Minister indicated how important environmental matters had become on the national agenda; but the rise of the environmental movement and 'green politics' in Australia has deep roots (Frawley 1999).

Heathcote (1975) has fitted the European settlement of Australia to 'five evolving human visions … as a result of the super-imposition of European and Anglo-Australian ideas onto those of the indigenous folk'; these visions being 'scientific, romantic, colonial, national and ecological' (p. 210). His final *ecological vision* encompassed parallel concerns about long-term productivity and the noneconomic ecological impacts of land clearance which in the late twentieth century, through a range of publications and the 1965 founding of the Australian Conservation Foundation (ACF), sufficiently raised public awareness to the extent that the 1972 test case over clearance of the mallee scrubs in Victoria's *Little Desert* established

[1] Situations and events in the natural world that lead to good or poor harvests.

[2] The arrangements and occurrences among people that determine the distribution of costs and benefits that flow from the operation of the food system.

that for the first time in Australia, the ecological view had triumphed over the colonial view: the desert was of more value in its natural state than under the plough, and environmentalism now ranked high among the political issues that decide the outcome of elections and the fate of governments (Sarokin and Schulkin, 1993).

Heathcote describes how this environmental vision grew out of, and contained components of, three out of his four earlier-described visions:

1. The *scientific vision* – derived from the eighteenth-century European scientific expeditions and their spirit of enquiry into natural phenomenon for their own sake
2. The *romantic vision* – which parallels the scientific vision but included a sympathetic response to the Aborigines and a delight in the 'uncivilised' nature of the landscape, along with a retreat or escape from the civilised environment
3. The excluded *colonial vision* – indifferent to the two prior visions and characterised by dissatisfaction with the natural scene and its lack of commercial opportunities from native flora and fauna (a body of opinion developed that saw the landscape as attractive only insofar as it was 'improved' by European man and his works – a vision with strong aesthetic and social implications)
4. The *national vision* built on the confidence and success achieved in the new landscape during the second half of the nineteenth century, which saw the realisation of the hopes of colonial developers as wealth accumulated from mines, field and paddock

Heathcote also describes how a major national task of the national vision was settlement of the tropical north, but while development attempts showed virtually unlimited optimism in the land resources, this optimism was not borne out by actual settlement. At a national symposium on the problem (of northern development) in 1954, the chairman of the Australian Institute of Political Science noted the general opinion that

> Australia could not justify her retention [of Northern Australia] unless she exploited to the full its mineral resources and its capacity for food production, and that our failure in this part of the continent seemed a national reproach which we should do our utmost to remove. (AIPS 1954, p. xiii)

Clearly at stake was not only the reputation of science but of the nation; but this view was in many aspects at odds with the emerging *environmental vision* for Australia.

In the 1994 revised edition of his book, Heathcote added a sixth vision – the *vision of guilt* – as previously rejected, ignored or unrecognised Aboriginal conceptions of the land began to permeate the European Australian consciousness with the realisation that an ancient culture had modified and sustained the land for tens of thousands of years prior to their settlement, a concept thoroughly researched and substantiated in Bill Gammage's 2011 book *The Biggest Estate on Earth: How Aborigines Made Australia.* This broader acceptance of the relationship between people and their environment was founded on 'certain beliefs which may not be explicitly argued, but are absorbed through a cultural framework which itself is being constantly remade by the outcomes of the relationship' (Frawley 1999, p. 265) and is a vision at odds with the historic and persistent western intellectual tradition of humanity standing apart from the rest of the animal world and nature in general. Unfortunately, this perspective is often popularly (and simplistically) associated with farmers per se – a technocentric or developmentalist viewpoint – whereas environmentalism is popularly identified as the perspective of 'the well-educated section of the middle class that do not draw their livelihood from the industrial or commercial sectors of the economy' (Frawley 1999, p. 289).

In reality, and it should come as no surprise, rural communities and agricultural land managers are generally very familiar with their landscapes including the reality of degraded landscapes and their associated costs and impacts, such as feral animals, weed invasion and declining biodiversity, though they might struggle with how to address these problems. There has been a wave of social and attitudinal change since the 1960s in regional Australia and particularly within agriculture as evidenced by emergence of the Landcare movement in the 1990s, widespread development and adoption of industry's best management practice (BMP) programmes and establishment of the regional natural resource management bodies network[3] in the twenty-first century. Today many non-Aboriginal Australians proudly and confidently talk about their 'sense of place' in the landscape (see Lankester 2013), and a majority of people I have worked with and interviewed throughout my career have described their affinity with and commitment to specific country. Evolving from Heathcote's environmental vision, *sustainability* (despite its difficulties in definition) has become a key concept for both pragmatic environmentalism and productive industry.

4.3 Ecologically Sustainable Development

The concept of ecologically sustainable development (ESD) came to the fore in the 1990s and enabled specific recognition and consideration of the interrelationships between social, economic and environmental aspects of development. The concept further expanded in the 2000s to include cultural considerations, particularly by Aboriginal and Torres Strait Islander people. ESD developed in part because of the 1970–1980s environmental focus on and critique of development ideology being countered by a backlash in the 1990s against environmental campaigns which threatened employment. The concept also appealed to governments trying to chart a course through the often-contradictory messages coming from the community. Now, environmental concern and activism has become commonplace in the world, prevalent in poor as well as rich nations, and with foci encompassing the local to global levels (Gallagher and Weinthal 2012).

This development and time-flow of perspectives in Australia was schematically presented by Frawley in Fig. 4.1, though this should not be interpreted as an end

[3]There are 56 Regional NRM bodies across Australia – see http://nrmregionsaustralia.com.au/regional-nrm/.

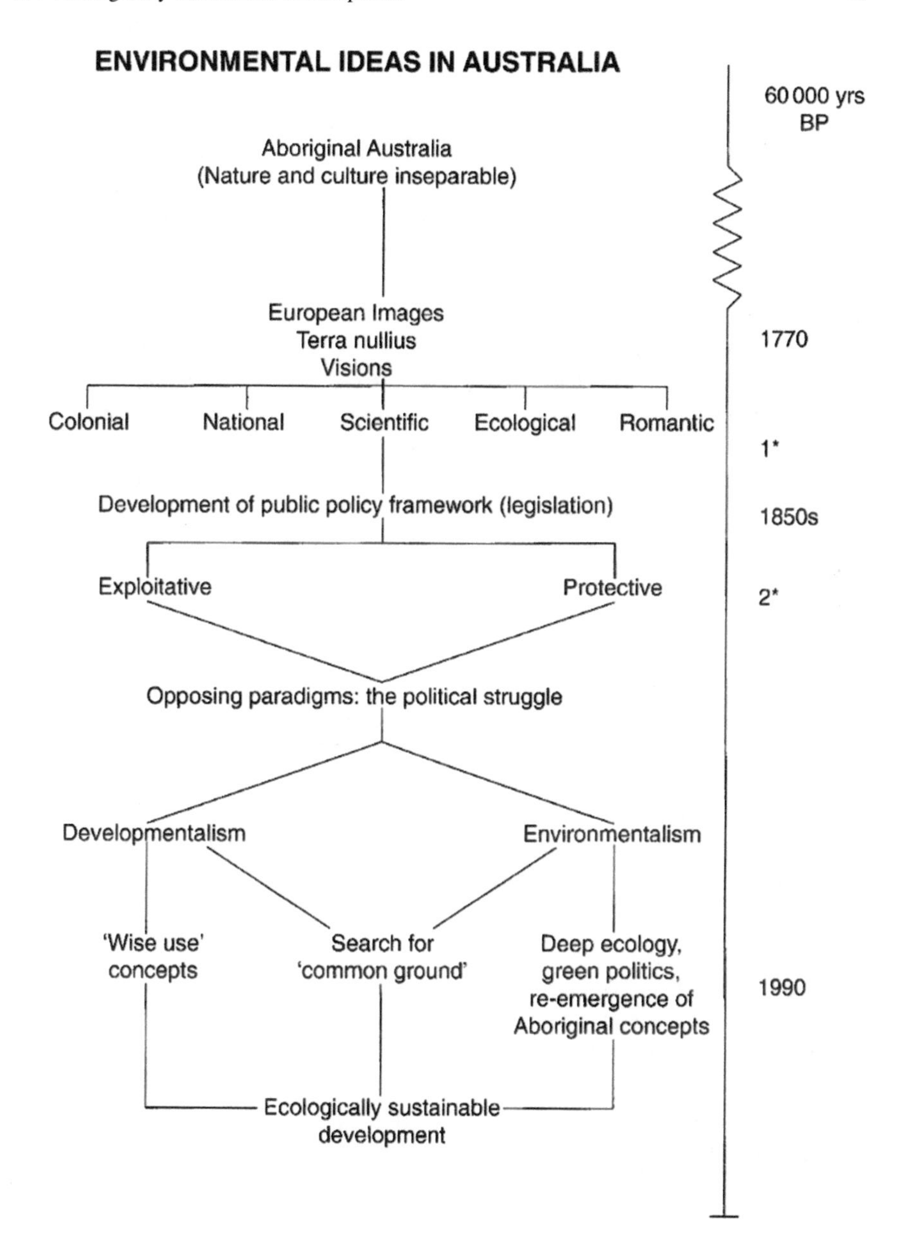

Fig. 4.1 Structure of environmental ideas and public policy development, Aboriginal Australia to present. (Source Frawley (1999) p. 267. Reproduced with permission from Kay Anderson and Fay Galeas, *Cultural Geographies 2nd Edition 1999 ©, Pearson Australia, page 267*)

point in ideas and policy development as there are problems with the concept of sustainable development.

The ESD concept can be deliberately vague and slippery, making it difficult to operationalise; it can enable *greenwashing* or *green camouflaging* of strategies and thereby foster hypocrisy; and it can imply the notion and possibility of sustainable growth, which Daly (1991) refers to as an impossibility theorem and therefore delusional (Brown 2011). The concept can also distract attention from other meaningful and perhaps more profound change, along with the root causes of global inequities and environmental degradation.

Frawley also considered the juxtaposition between the protective and exploitive components of environmental law, where the protective component aims to protect the environment from undue degradation by human activity, and the exploitive component governs the disposal (sale, lease, granting) of natural resources. This segregation can result in a situation described by Lockie (2014) as 'ecological apartheid' – where areas are either 'locked up' or 'trashed', with nothing in between. While such an outcome has most likely never been a deliberate policy intention, there are many instances where this juxtaposition is *perceived* to occur, particularly at the production/protected area interface; for example:

- 'Australian politicians and conservationists have often commented (off the record) that farmers' attitudes to the environment are not conducive to effective land management' (Vanclay and Lawrence 1995).
- A banana grower and pastoralist – 'I have no [feral] pigs living on my farm - they live in the national park and trespass on my farm every night' (Noble, 1997, p. 83).
- '[The] implication [being] that the local land occupiers could not be trusted to preserve the environment' (Mather, 1993).

Such perceptions easily translate into polarised stereotypes or attitudes that can actively inhibit collaborative outcomes. These polarised attitudes have, in part, been attributed to Northern Australia being viewed as an 'over-managed and under-led last frontier where progress requires indigeny [sic] to be assimilated and hostile environments conquered, and where wealth comes from beneath ground or off the hoof rather than between the ears' (Ellyard 2004, p. 2).

Ellyard advocates that for Northern Australia to participate effectively in the twenty-first-century economy, much more than provision of improved infrastructure is required: it will need a *new frontier* mentality, where design and innovation result in doing more new things first, rather than better ways to do old things. Many leading industries have discovered and benefited from opportunities for improved processes that increase productivity by embracing interrelationships between social, economic and environmental elements of ecologically sustainable design, and agriculture should not fear learning from and aligning with such expertise.

At the same time, there is a growing perception and understanding in society and among decision-makers that to properly address environmental problems will require fundamentally new approaches that encompass proper consideration of the aspirations and perspectives of all concerned members of both local and international communities – that is, applying the insights from environmental humanities to environmental problem-solving, because 'environmental problems are inherently human problems' (Kueffer et al. 2017, p. 3).

4.4 Capital Ideas

As the concept of sustainable development developed traction and understanding within the broader community, an observation spread within the agricultural sector was that *it's hard to be green when you're in the red*. This comment described the need for farmers to maintain economic solvency if they were to have any hope of satisfying a broader community expectation for higher standards of environmental stewardship. Financial capital is only one form of the material and human resource assets that society can capitalise though, for as Bourdieu (1986) explained, the term *capital* can refer to any resource or asset that social actors can employ to further their goals – whether they be economic, cultural or social. Bourdieu argues that it is the distribution of the different forms of capital that represent the 'the immanent structure of the social world, i.e., the set of constraints, inscribed in the very reality of that world, which govern its functioning in a durable way, determining the chances of success for practices' (p.241).

Coleman (1988, p. S100) explains how in the same way physical capital is created by changes in material to form tools, 'human capital is created by changes in persons that bring about skills and capabilities that make them able to act in new ways'; but Coleman also draws an important distinction between human and social capital because 'social capital comes about through changes in the relations among persons that facilitate action' (emphasis added). Stokols et al. (2013) provide an overview of these two asset/capital classes (see Table 4.1) and propose that this

Table 4.1 The evolving forms of capital

Material capital	Human capital
Economic capital or material goods that facilitate creation of new products and financial growth	Human capital, created through changes in persons that equip them with new skills and capabilities that enable them to act in new ways
Natural capital or those resources produced through nature-based processes and include all privately held ecological and biological assets	Social capital or changes in the relationships among persons that facilitate their coordinated action for mutual benefit
Human-made environmental capital, such as buildings, vehicles, tools and other products created by people	Moral capital or the investment of personal and collective resources towards the cultivation of virtue and justice
Technological capital, a subcategory of human-made environmental capital exemplified by telephone systems, computing equipment and Internet-based services	Social media capital or the value which is developed by the connections, the reputation and the influence of an individual or an organisation, in the social media environment

multifaceted conceptualisation of capitalised assets has direct relevance for resilience theory because it:

> highlights transactions wherein decrement in one form of capital (e.g., hazards from extreme weather events) are addressed through the mobilization of other forms of capital (e.g., social capital in terms of a network of emergency service providers; moral capital in the form of norms about sharing in times of need). (p.5)

In Australia, there has been considerable debate since the 1990s over the significance of social capital for community development and policy formulation (Giorgas 2007) as it 'encompasses norms and social networks which facilitate social action, thus enabling individuals to act collectively' (p. 207). The concept was further refined by Putnam (2000) who 'has helped place social capital as a central policy concern for local, state and federal governments, as well as international organisations such as the World Bank' (Giorgas 2007, p. 209); and for whom networks, norms and trust are central, observing that '... trustworthiness lubricates social life' (Putnam 2000, p. 21). Stehlik (2010, p. 93) states that 'the critical factor in social capital building is trust', and Hanna et al. (2009, p. 31) describe social capital as the 'features of social organisation, such as trust, cultural norms and networks by which communities facilitate action or simply keep themselves going'. When investigating the reluctance of natural resource users to adopt seasonal climate forecasts that could enhance their resilience, Marshall et al. (2011) found that social factors were more important than technical ones in explaining their reticence, not surprising, as farmers are human and humans are social beings.

Putnam further described and emphasised two fundamental aspects of social capital: *bonding* and *bridging* social capital. *Bonding social capital* relates to the relations among relatively homogenous groups of people which can promote exclusiveness, build social walls and be less tolerant of diversity. *Bridging social capital* on the other hand involves the building of connections between heterogeneous groups, which can foster links to external assets and generate greater information diffusion. Putnam argued that both forms of social capital are necessary and in the right mix create consensus and mutual obligation, which contributes to economic prosperity and effective governance.

Another conceptual refinement introduced to social capital literature is *linking social capital.* This concept refers to the extent to which individuals build relationships with institutions and individuals who have relative power over them, for example, to provide access to services, jobs or resources (Hawkins and Maurer 2010; Szreter and Woolcock 2004). Linking social capital is most clearly connected with structural approaches to power and is central to an understanding of how bonds and bridges enable collective agency within community (Onyx et al. 2007). This concept includes *social media capital*, which relates to the currency of reputation that influences online business success and influence. While everyday communication is a primary contributor to the accrual of all forms of social capital, the use of social media to express care and concern for others is increasingly important to social capital outcomes (Quinn 2016).

Social capital has gained intellectual currency as a means to understand the relative strength of families and communities, and Hawkins and Maurer (2010) state that its focus on the actions of the individual in relation to their community is where the concept is of greatest value, concurring with Schuller et al. (2000) that social capital is unique in its

> ability to bridge the theoretical gap between individual and community that spans from the micro to the macro in an interactive and independent manner more effectively than many previous socio-economic/-political theories. (p.1779)

The devil is always in the detail though, so whether social capital is used to empower or disempower will depend on 'the particular intersection of social capital and power relations within specific rural networks' (Onyx et al. 2007, p. 289) and will always need to be considered and understood.

4.5 Traditional Owners and Agriculture

Bill Gammage's comprehensively researched 2011 book *The Biggest Estate on Earth compares* the diary notes, descriptions, illustrations and paintings of the early European explorers of Australia with more recent photographs and descriptions to substantiate his thesis that pre-European (1788) Australia was an intensively human-managed landscape – that is, an ecologically based spatial and human arrangement. Gammage describes how early Europeans repeatedly commented that the land looked like an estate or a park, which

> hardly reflects people constantly on the edge of want. They cannot have been the scavenging, chance-dominated savages Europeans thought them. A rich and time-eating spiritual life builds on abundance, not poverty. In the driest and most fire-prone continent on Earth, abundance was not natural. It was made by skilled, detailed and provident management of country. (Gammage 2011, p. 138)

Park-like it may have seemed, but this Aboriginal system of land management did not look like farming as they knew it to early European settlers – 'People farmed in 1788, but were not farmers' in the European sense (Gammage 2011, p. 281). This is a very important distinction, as to this day '[many] Europeans think farming explains the lifestyle differences between them and Aborigines' despite the ample evidence now provided by Gammage that pre-1788, agriculture was 'spread more widely over Australia than now' (p. 289).

Post-1788 agricultural engagement with Aboriginal people was characterised initially by their dispossession and persecution, before moving through a period of exploitation prior to today's widespread disengagement. This has occurred despite the extensive influence of Aboriginal culture on early settlers, particularly in the cattle industry, which still draws heavily on language, fire management and understanding of natural signs and processes (personal observation and Gammage (2011, p. 167)). Pascoe (2014) also presents historical information indicating many areas of pre-European Australia could have been an agricultural society, though again not of a form recognised as such by early explorers; and Pascoe hypothesises that this myopia could have been deliberate, as 'the belief that Aboriginal people were "mere" hunter-gatherers has been used as a political tool to justify dispossession' (p. 129).

This impact of immigrant farmers on Indigenous land managers is not confined to Australia, and the late twentieth-century re-emergence of Australian Aboriginal land management concepts within the *Environmentalism* paradigm described by Frawley (1999) is also being reflected around the world: de Sartre and Taravella (2009) describe how internationalisation of the Brazilian Amazon combined with sustainable development aspirations is changing the governance of modern states, particularly in how they exercise their sovereignty – recognising local to global spatial scales, stimulating emergence of a hybrid form of governance through public-private partnerships and giving importance to local stakeholders including Traditional Owners who are not usual policy considerations.

It is important to recognise though that while acknowledging relationship is important for mainstream Australia's recognition of its First People's continuing interest and involvement in their land and its management, contemporary Aboriginal land management aspirations are not limited to achieving environmental outcomes. Indigenous Northern Australians comprise an average of 15% of the population of Northern Australia (compared to 2.3% in Southern Australia), with much larger Indigenous representation in the populations of northern Western Australia and Northern Territory (approximately 25%) and in populations outside of the main urban centres across Northern Australia (well in excess of 50% in many instances). As a result, Indigenous organisations are a key component of the government, industry and non-governmental organisation's institutional framework in Northern Australia. Northern Australian Indigenous interests in the land and sea estate are much greater than they are in Southern Australia, with Indigenous interests owning or exercising some degree of legal control over close to 80% of the Northern Australian landmass and considerable areas of sea country.

Indigenous Australians are actively addressing their social issues and considering alternate futures and meaningful employment opportunities for their growing communities, along with the complexity of individual land ownership, entitlement and responsibilities. Native title is limited (and sometimes limiting) in its ability to address these issues, and there are many Indigenous people in Northern Australia who are not Traditional Owners and therefore unable to directly benefit from enterprise that is derived from land, water and sea interests and rights. Furthermore, there are many actual and aspiring Indigenous-owned and Indigenous-operated businesses in Northern Australia that do not have any link to land, water or sea interests and rights but still face the same generic structural challenges associated with Northern Australia, as well as the unique challenges faced by Northern Australian Indigenous-owned and Indigenous-operated business. Any strategies for northern development must be cognisant of both this complexity and the opportunities to achieve a better outcome for all.

Of equal, if not more important, during the tens of thousands of years, Northern Australia has been home to Indigenous Australians (Bottoms 2013); there have been sea level fall and rise, extensive volcanism and many cyclones and other natural events commonly referred to as natural disasters; and it is more than probable that within traditional ecological knowledge systems, there is lived experience that would assist Australia's more recent arrivals in their relationship to this country, if we choose to do so.

4.6 Climate Change and Other Disasters

The *planet and people matters* previously described by Farmar-Bowers (2011) can change in regular, predictable patterns in response to seasonal cycles and social rhythms or suddenly and dramatically as a consequence of unforeseen or cataclysmic events. Such events are commonly referred to as disasters. Disasters take from people the things they hold dear and require them to come to terms with dramatically changed lives, and Australia has long been seen as a land regularly challenged by natural disasters. While Australia would not commonly be considered an overly 'poetic' nation, many of a 'certain vintage' know by heart these lines from Dorothea Mackellar's (1908) poem *My Country* (originally published as *Core of my Heart*):

> I love a sunburnt country,
> A land of sweeping plains,
> Of ragged mountain ranges,
> Of droughts and flooding rains.
> I love her far horizons,
> I love her jewel-sea,
> Her beauty and her terror
> The wide brown land for me!

Disasters are multifaceted, and their impacts are not constrained within a single aspect of an individual's or a community's life. They are the convergence of hazards with vulnerabilities, and, as such, any increase in physical, social, economic or environmental vulnerabilities through such factors as population growth, change in land uses or climate change can contribute to an increase in the occurrence of disasters (Guha-Sapir et al. 2013a; McDonald 2013). During the last 50 years, the better reporting of disasters combined with population growth, growing population density in risky areas and changes in land use has led to an increasing number of natural disasters being reported worldwide, although 'poverty remains the main risk factor determining the long-term impact of natural hazards' (Guha-Sapir et al. 2013b, p. 1). Natural disasters have a tremendous impact on the poorest of the poor, who are often ill-equipped to deal with natural hazards and for whom a hurricane, an earthquake or a drought can mean a permanent submersion into poverty (Guha-Sapir et al. 2013b).

This increase in occurrence of disasters is not equally distributed among continents: 'Asia and Africa show the greatest growth, the Americas and Europe show a similar evolution, whereas disaster occurrence in Oceania remains relatively stable over time' (Guha-Sapir et al. 2013a, p. 9). But there is wisdom in Milton Friedman's[4] observation that 'only a crisis – actual or perceived – produces real change. When that crisis occurs, the actions that are taken depend on the ideas that are lying around'.

It is in relation to natural disasters that the word resilience is most often used by Australia's news media, and many see this ongoing struggle with natural disasters as a fundamental shaping force on our national identity and psyche – a 'part of everyday living in much of northern Australia' (Pickup 1978, p. iii). Furthermore, during the early European settlement of Australia 'most inhabitants regarded the ups and downs of climate as being an unfair imposition from on-high, rather than a normal feature of the environment, to be managed and celebrated' (Stafford-Smith 2005, p. 5); but this attitude has changed as the highly variable nature of Australia's climate is better understood, an understanding assisted by the realisation that 'many Australian species may have evolved to "use" fire' (p. 7).

Notwithstanding these evolving perspectives, natural disasters are a real phenomenon in Australia's landscape. A 2015 Deloitte Access Economics report by the *Australian Business Roundtable for Disaster Resilience and Safer Communities* included the first analysis of the economic costs of the social impacts of natural disasters – they cost the economy more than physical impacts like property damage! In 2015 the economic costs of natural disasters (tangible and intangible) exceeded $9 billion, which was equivalent to 0.6% of gross domestic product (GDP) in the same year. This figure is expected to almost double by 2030 and to average $33 billion per annum by 2050 in real terms, even without considering the impact of climate change. A separate report estimates that $17 billion (in present value terms) will need to be spent on the direct replacement of essential infrastructure between 2015 and 2050 through natural disaster damage (Deloitte Access Economics 2016).

Human-induced climate change is increasing the risk of both extreme weather and temperatures in Australia, and the CSIRO (2014) *State of the Climate 2014* reports Australia is now almost a degree warmer than a century ago, with:

- Seven of Australia's 10 warmest years since 1998
- Very warm months at five times the long-term average over the past 15 years, while very cool months have declined by a third
- Winter rainfall declined 17% since 1970 in Australia's south-west and by 15% since the 1990s in the south-east
- Increased extreme fire weather and a longer fire season since the 1970s
- Sea-level rise likely to increase the frequency of extreme sea-level events
- Tropical cyclones decreasing in frequency but increasing in severity
- By 2070, depending on how fast greenhouse gas emissions are reduced, temperatures between 1 and 5 °C warmer than the 1980–1999 average

[4] An American economist who received the 1976 Nobel Memorial Prize in Economic Sciences for his research on consumption analysis, monetary history, and theory and the complexity of stabilisation policy, in his 1962 book *Capitalism and Freedom.*

Natural resource-based communities such as agriculture are generally viewed as vulnerable to the risks and disasters which will be affected by climate change (Flint and Luloff 2005), so environmental hazards will continue to be part of Northern Australia life. Every year, communities are threatened by cyclone, flooding, drought or bushfire, while farmers additionally face pest and disease outbreaks which are also often climate-related.

While the idea of disaster usually conjures images of the spectacular – 'the catastrophic failure of technological systems, the devastation of natural disasters, the creeping threat of global warming etc.' (Lockie and Measham 2012) – farmers are also at risk from legislative change that can impact their business overnight (e.g. the aforementioned 2011 live cattle export ban). This tragedy however assumed Shakespearian[5] proportions particularly because of the sense of deception felt by cattle producers around the government's reaction to the *4 Corners* television story.

In Australian agriculture, the neo-liberal state has sought to foster self-reliance in the management of risks, with a particular focus on drought. Drought was once seen as a national problem to be addressed through state financial support, whereas now it is more often reconstituted as a problem for individual farmers to deal with as part of rationally based risk management (Cheshire and Lawrence 2005). From 1971 to 1989, farmers were provided with disaster relief at times of severe drought, but now state policy recognises drought as a normal feature of the Australian landscape and provides struggling farmers with various types of business support assistance, with income support (not drought relief funds) provided only in exceptional circumstances (Aslin and Russell 2008). This has occurred at the same time as other government supports, particularly industry extension services, have been significantly reduced through neo-liberal reforms (Gill 2011).

Under Australia's federal political system, the responsibility for natural resources is vested in the individual states, which consequently replicate bureaucratic structures to implement laws for fundamentally similar purposes in each state (Stafford-Smith 2005). However, the Australian Government can provide funding through the Natural Disaster Relief and Recovery Arrangements (NDRRA) to help pay for natural disaster relief and recovery costs, and a state or territory may claim NDRRA funding if (1) a natural disaster occurs; (2) state or territory relief and recovery expenditure for that event exceed agreed thresholds; and (3) the state or territory notifies the Attorney-General's Department of the event. NDRRA applies to the following natural disasters, bushfire, earthquake, flood, storm, cyclone, storm surge, landslide, tsunami, meteorite strike and tornado, but does not apply to drought, frost, heatwave and epidemic or disaster events resulting from poor environmental planning, commercial development or personal intervention (other than arson) (Australian Government 2014).

Drought was removed from the NDRRA in 1989 after a review by the Drought Policy Review Taskforce specifically stressed that drought was a natural, recurring and endemic feature of the Australian environment, and the prospect of variable

[5] Othello, for example, is a tragedy that proceeds from misunderstandings and miscommunication.

seasonal conditions was a normal commercial risk that must be incorporated into the management of Australian rural enterprises. Sustainable farm management was supposed to be able to withstand drought impacts without official relief efforts (Heathcote 2002). For comparison, since 1988 the USA has spent US$48 billion on drought assistance rather than on mitigation systems which would have been cheaper (Wilhite and Pulwarty 2005, p. 396), though America's climate does not have the renowned variability of Australia.

However, by the time drought becomes a mainstream issue resulting in a political imperative to do something, it is usually at a late stage where emergency measures are required for those individuals in desperate need; and despite its banishment from the official ranks of natural disasters, drought still makes powerful news headlines, tugs at public sensitivities and drives political decisions: Commonwealth drought support measures were $21 million in 1993, $42 million in 1994, $147 million in 1995, and $211 million in 1996 (O'Meagher et al. 2000). The 2016 budget provided $250 million to continue the Drought Concessional Loan Scheme for an additional year and $25 million to assist farmers reduce the impact of pest animals in drought-affected areas; and in 2018 the Minister for Agriculture proudly stated, since taking office in 2013 the government had committed more than $1 billion to drought assistance through farmhouse assistance, concessional loans and a community drought programme (Schwartz 2018). At the same time, farmers were also supported by numerous charitable and private initiatives including *Rural Aid*, *Thirsty Cow*, *Drought Angels* and the *Salvation Army*, which indicates the continuing public sympathy for farmers *doing it tough.*

It seems that only 'a minority of managers apply for drought relief and of those, a further minority get most of the relief' (Heathcote 2002, p. 22), and the Australian Farm Institute recommends that avoiding drought-induced farm business financial crisis must be the aim of responsible government policy (Heath 2018) perhaps because 'a drought hazard is produced as much by a human inclination to bank upon the average or better conditions occurring as it is by deficient rainfall' (Oliver 1980, p. 16) or, more probably, that the focus on risk places drought in the economic sphere rather than the social or emotional spheres and thereby avoids the 'recognised fact that for many, farming is a way of life, rather than an economic undertaking' (Stehlik 2005, p. 68).

Move away from the media headlines though, and it is evident that increasingly Australian farmers realise Australia's climate reality. Charles Massy describes the day he

> realise[d] that the predicament [drought] I was in was no one's fault but my own. The reality of Australia is that it has regular droughts. The land, soils, microorganisms and other creatures and vegetation are adapted to this. And when irregular rain comes, it is often heavy and brief. To manage this 'normal' Australian occurrence requires a completely different mindset, and a different management approach and techniques to those of northern-hemisphere agriculture. (2017, pp. 292–293)

This focus on drought has been included for two reasons: Australia's irregular rainfall will continue to be a significant influence on Northern Australian agriculture; and of more importance, it demonstrates that the paradigmatic shift from disaster response to risk management has not been universally accepted, even by government; and in reality

> as Australian agro-ecosystems generally have to cope with high levels of climatic variability at all scales … policy intervention needs to be aware of the many different local ways in which climatic variability plays out and affects managers across the continent, [and therefore] drought simply brings some underlying critical structural problems to a head. (Stafford-Smith 2005, p. 12)

The antithesis of droughts is floods, and one often follows the other regular as the lines in Dorothea Mackellar's poem. Often these floods are a consequence of tropical cyclones (King 2001; Nott and Price 1999), and these biophysical impacts have direct and indirect effects on the health and wellbeing of people living in affected regions (Green 2006).

Fire is another component of the Northern Australian landscape, with its importance increasingly understood by scientists informed by Indigenous Australians. Fire also provides opportunities for carbon management (Douglass et al. 2011; Fache and Moizo 2015; Maru et al. 2014; Price 2015; Price et al. 2014; Richards et al. 2011; Richards et al. 2012; Russell-Smith et al. 2013; Stafford-Smith 2005; Whitehead et al. 2008; Wilman 2015). The wet-summer cycle means communities are not as vulnerable to wildfire as those located in Southern Australia's dry summer-dominated areas, and controlled burning enables 'efficient mitigation of wildfire [which] contrasts markedly with observations reported from temperate fire-prone forested systems' (Price et al. 2012, p. 297); though Cottrell (2005, p. 109) reminds us that 'locality remains important to the understanding of communities, bushfire hazard and delivery of services'.

In recent times the 'easy availability of large databases and information systems has played a significant role in prompting analyses of community vulnerability' to disaster events (King 2001, p. 147), though King warns of a danger in defining and measuring vulnerability 'because they are there, rather than because these databases encapsulate vulnerability'. As he points out, a 'large population in a hazardous location alone defines maximum vulnerability. All other measures modify that basic classification' – important to remember when the development of Northern Australia is being actively promoted.

Most industries like certainty of tenure to plan, which brings us to risks to agriculture associated with an expansion of extractive industries across Australia. While there is uncertainty around the current and future economic, social and environmental impacts of the mineral and energy industries, these uncertainties are sometimes unnecessarily exacerbated by the existence of separate and often unrelated legislative frameworks governing mining (Mayere and Donehue 2013). Often, agriculture operates under the assumption that existing land users will be entitled to continue into the foreseeable future, but Mayere and Donehue (2013, p. 221) found that while distrust in plans and policies was strongly related to an imperfect knowledge of land use and tenure, it was the 'state's capacity to override supposedly comprehensive local planning instruments [that] lies at the very core of the uncertainty' and thus present an additional risk element.

4.7 Australian Agricultural Policy in the Anthropocene

The interplay around sectoral aspirations for an intensification of production agriculture in one of the world's oldest contiguously managed human landscapes is occurring at a time when the extent of humanity's true impact on the Earth through population increases and anthropogenic emissions of carbon dioxide are being interpreted as a new geological epoch – the Anthropocene (Crutzen 2002) – and at a time when decision-makers have the knowledge and understanding of their responsibility to manage humanity's relationship with the planet.

While climate change has provided the focus for demonstrating human's ability to influence Earth's environment *as a single, evolving planetary system* (Steffen et al. 2011a, p. 842), additional impacts include changes to the nitrogen, phosphorus and sulphur cycles fundamental to life, altering water cycles between the land and atmosphere and likely driving the sixth major extinction event in Earth history. Given the magnitude and implications of these changes, it is hardly surprising that global society is taking collective and rapidly increasing interest in the ramifications of associated policy, particularly in achieving security for future generations. Sustainability has become a key concept, having proven itself a robust tenet around which dialogue and mutually agreed action can be developed between industries and the conservation sector (Taylor 2010), though Rowell (1996) argues that this 'accommodating' mode is actually the co-opting of the agenda for change by dominant political and economic interests.

Though the political influence of environmentalism in Australia may have waned (temporarily) in the aftermath of the 2007–2008 Global Financial Crisis (GFC) and changes of government at both state and federal levels, modern environmentalism as the manifestation of a social and political movement advocating a new (or arguably rediscovered) philosophy of human conduct towards nature and the cultural artefacts of human civilisation and other human beings will continue to challenge more exploitive attitudes and industries. As Lockie (2004, p. 26) points out, 'social theory that cannot find a place for the non-human organisms, substances and patterns of nature is social theory that is inadequate for understanding key dimensions of our contemporary world'. Individuals and institutions are demanding the right to participate in development decision-making and governance, and as the focus of development shifts from modernisation to sustainability, the respective roles of government and individuals in systems of governance are also shifting. Eversole and Martin (2005 p. 1) state that 'Good development policy and practice must do more than ensure economic efficiency – they must promote human well-being and equity, and this is not a task that states or development agencies can carry out on their own' (see also de Sartre and Taravella (2009).

This new environmentalism paradigm embodies a continuum of thought and action concerned with the relationship between people and their environment which can manifest in extreme ecocentric views, for example, some elements of the animal rights movement (Singer 1990), and their ability to directly affect established industries was demonstrated by the 2011 federal government ban on live cattle export to

Indonesia. This, and similar events in this rapidly evolving era of instant global communication and social networks, leads one to consider Habermas's (2010, p. 192) assertion that 'Debate and dispute in the "weak" publics of the press and the street are better suited for the "struggle over needs" and their interpretation than it is for the resolution of these conflicts'. While the extensive media coverage and debate did not resolve the conflict, neither did the subsequent political decision to ban exports, though the impact of the decision continued to be felt in Australis's cattle industry 5 years after the event. It was, to use Habermas's term, the *weak* media debate that generated the policy decision.[6]

Whether weak or strong, 'as economic prosperity rises, public concern for the welfare of animals increases throughout the developed world ... a trend likely to continue within the growing middle classes of the BRIC economies [Brazil, Russia, India and China]' (RSPCA[7] Australia (2014, p. 40) in an animal welfare-themed edition of the Australian Farm Institute journal). While the approach industry takes in responding to public concerns can limit the extent of societal conflict and the level of impact upon business and create new market opportunities, animal welfare is a public good – which means it is the responsibility of everyone to improve it, not only farmers. However, consider that massive changes introduced to the British pork industry in 1991, in part related to implementation of new animal welfare standards, contributed to weakened farm productivity and profitability to the extent that 20 years later, while sow stalls had disappeared from British farms, so had UK pork production, with a majority of product now imported from nations without similar animal welfare standards. The problem just shifted from the public view (Sloyan 2014).

Animal welfare is an example of how a minority can impact an industry or region, but the converse can also be problematic – where a majority impinges on the rights of a minority. Dale (2014, p. 8) describes a north-south cultural divide for Northern Australia, where 'waves of different social, political and economic agendas have washed over the north' from the developed south. As an example, he describes 'resource preservationists who seek a northern wilderness but appear to lack much empathy for those people who actually live in, care for, and derive an income from the northern Australian landscape'. Herein lies a conundrum: how, in a democracy like Australia, to respect and protect the views of the less than 5% of the national population who actually reside in the more than 40% of land that is tropical Australia?

The question is even more complex: the popular view of northern life has been strongly influenced by our historic European perspective[8] of the tropics as a place of isolation, hardship and extremes, whereas the tropics are 'just home' to 40% of the world's population and 55% of the world's children under 5 years of age. Also, the

[6]There exists a view within the pastoral industry that the media debate was also weak intellectually, insofar as there was only a limited effort to research ramifications of the decision or to present an industry perspective.

[7]Royal Society for the Prevention of Cruelty to Animals.

[8]Aristotle described the tropics as the 'torrid' zone, too hot for habitation.

tropical economy is growing 20% faster than the rest of the world, and the tropics host 80% of terrestrial biodiversity (JCU 2014). Northern Australia, as a sparsely populated tropical region, is a global aberration and, as such, needs to be conscious of an overly parochial view, particularly in this globally connected era.

Dale et al. (2014) also provides a detailed account of modern north Australian history, not for the sake of record keeping but 'to progress a national dialogue about a cohesive forward agenda' (p. 9), and builds a strong argument for adopting region-centred governance systems capable of integrating improved natural resource management and economic and social development. Such an approach would benefit our globally connected world, particularly tropical regions, many of which were until very recently managed by first-world European countries. Improved connectivity and communication are hallmarks of our era and, despite the risks of submerging policy-makers in too much information, provide the opportunity and tools for informed cross-community, regional and national collective governance. If Australia can achieve this transition from government to governance for the important and (rightly) contested decisions regarding the agricultural development of Northern Australia, the opportunity is there to extend the process for global benefit.

4.8 Global Decision-Making and Evolving Governance

Governing through knowledge instead of politics is a century-old utopian dream, inspired by a sense that politics is messy, irrational, selfish and short-sighted. However, 'reasoned analysis is necessarily political' (Stone, 2012, p. 375). Reason doesn't start with a clean slate on which our brains record their pure observations. Reason proceeds from choices to notice some things but not others, to include some things and exclude others and to view the world in a particular way when other visions are possible. Policy analysis is political argument and vice versa.

Though not a Northern Australian example, the dichotomy in rural policy-making is demonstrated by responses to the escalating and often disastrous bushfires of southern Australia: Stephens (2010, p. 18) identifies what he believes to be an increasing trend towards 'megafires' over recent decades and advocates 'political and community commitment to a landscape approach to fire risk management that looks at the risks and options across all land tenures and is supported by scientific evidence', whereas evidence shows that government focus was actually on high-profile investments in fire suppression rather than in adequate fire prevention measures (Keogh 2010). This divergence in focus is attributed to an inability (or unwillingness) to resolve pre-existing and conflicting priorities, such as a public discomfort with the smoke and ash associated with prescribed burning near densely settled areas, a lack of resources available to public land management agencies for proactive activities and a prevailing view among some sectors that fuel reduction (through either burning or grazing) produces undesirable environmental damage. The customary institutional tool for determining outcomes in such community conflict, particularly over property rights and access and use rights, has been land tenure and its coupled obligations; but the above example demonstrates that this mechanism cannot cope with multiple overlays of community expectation. Without community consensus to an agreed position, it is difficult to envisage resolution of such problems.

Returning to Northern Australia, until recent times pastoralism dominated the rangelands as the highest and best use. However, beginning in the 1980s, pastoralism has been steadily displaced as other interests including conservation, indigenous ownership, tourism and mining have received recognition. While these are not exclusive activities, Holmes (2000) suggests that an overextension of private property rights (driven by land speculation) has been, and continues to act as, a barrier to transfer of land to other tenures. There is an assumed right of exclusive possession (despite being subsequently disproved in the 1996 Wik High Court decision[9]) and an effective right of veto over excision of small parcels for private purposes, such as a roadhouse or intensive agriculture venture. Holmes argues that a process of *de-privatisation* is a logical response to changing resource values and expectations and would be consistent with the original intent of the lease tenure system to facilitate transfer of land to *higher* uses when circumstances were right. This multifunctionality concept, where agriculture is only one component of land use, could have a 'pivotal role in the reconstitution of rural space' (Holmes 2012, p. 252), though it should be remembered that it was misapplication of a very similar principle that resulted in the previously discussed difficulties of nineteenth-century pastoralists described by Bennett (1927).

Globally, Holmes (2000) describes this shift in resource values as a transition from a *productivist* era, in which commodity outputs were universally given priority, to a *post-productivist* era in which emerging amenity values are increasingly important:

> rural lands are undergoing a major re-evaluation. Productive lands, surplus to requirements for commodity outputs, are increasingly in demand for their amenity values, broadly defined as those values directly satisfying human needs and wants, other than material needs. In the [Australian] rangelands, these new values embrace Aboriginal traditional and contemporary uses, biodiversity conservation, and preservation of cultural and natural heritage, tourism and recreation, among others. (p. 135)

A powerful example of this post-productivist transition comes from Europe and North America, where serious (and often controversial) efforts are being made to reintroduce large native grazers and predators to areas where they have been locally extinct for millennia (Heffernan 2016). Often these areas are still farmland. Such effort would probably have seemed unwise at a time when these animals competed directly with humans for survival but is now a clear indication not only of humani-

[9] On 23 December 1996, the High Court handed down its decision in *Wik Peoples v State of Queensland and Others*. The decision confirmed that native title rights and interests may exist over land which is or has been subject to a pastoral lease and possibly some other forms of leasehold tenure. The Court held that existing pastoral leases issued prior to 1 January 1994 and the rights granted under them are valid. It also held that the rights of the pastoralist prevail over native title rights and interests to the extent of any inconsistency (ATNS database 2004).

ty's dominance of the natural environment but acceptance of the concomitant responsibility that accompanies this, at least by some.

This new, highly variegated geography of rural development, settlement and land use is producing many different *rurals*, each with varying developmental capacities and accompanied by shifts in political structures, agency and representation (Argent 2011). This change of definition carries a particular risk for contemporary Northern Australian agricultural development where a statement by a producer, politician or policy-maker is susceptible to very different interpretations by other community sectors, depending on their personal interpretation of the words used.

While care must be taken in extending the experience of the rural elsewhere to Northern Australia – as they are not necessarily directly comparable with the productivist multifunctional countryside transitions of the majority of advanced industrialised nations – global trends are at work and gathering momentum; and in democracies, voter numbers carry more weight than ownership. Argent (2011, p. 184) describes three key processes in this transition, one of which definitely applies to the Northern Australian context: the 'increasing public regulation of farm production techniques in order to prevent and/or control environmental and welfare outcomes. Of the other two processes, the 'withdrawal of longstanding support to the sector' (process 2) is not uniformly consistent with contemporary political and sectoral encouragement for the region's agricultural expansion; and process (3) 'in-migration into select rural areas that has produced a new 'class' of rural residents who have actively challenged the hegemony of farmers over the directions of local social and economic challenge' has certainly not occurred to the extent experienced in other advanced industrialised nations, a consequence of distance and Northern Australia's small population.

This does not mean Northern Australia is not subject to the influence of supranational single-interest groups such as People for the Ethical Treatment of Animals (PETA), rather their influence is generally seen to emanate from the southern capitals and not from sectors residing in the community and is therefore an extension of Dale's *North South Cultural Divide*. This in turn can result in entrenched philosophical positions rather than any attempt to understand another's perspective. However, the future of Northern Australia is inexorably bound to the expanding and overlapping mesh of global perspectives, networks and influences; and these include specific focus activist groups whether they care that their actions are fair, equitable or rational. While care must be taken in applying the experience of elsewhere to Northern Australia, there is probably more to be learnt from exploring similarities than from elaborating differences.

Returning to Argent's first key process – 'regulation of farm production … to … control environmental and welfare outcomes', Costanza et al. (2014) estimated the 2011 value of the services provided by ecosystems of fundamental importance to human wellbeing, health, livelihoods and survival is $125 trillion per year[10]. While

[10] A move away from gross domestic product (GDP) as a misleading measure of national success is also advocated, and adoption of a metric more suited to measuring what makes life worthwhile (Costanza et al. 2014).

this (verging on the impossible to comprehend) figure is actively used to argue against increasing productive development by some sectors, it is also used to argue strongly for improving the perilous financial position of many rural industries while bolstering a transition to more sustainable production. *Market-based instruments* (MBI) or *payments for ecosystem services*[11] (PES) have been promoted as economically efficient targeted solutions to otherwise intractable environmental policy problems, with the added potential to improve the livelihood security of ecosystem service providers (Lockie 2013). Here, aligned with the concept that it is 'unfair to expect [farmers] to manage to a standard of environmental sustainability that cannot practicably be achieved through environment-blind market forces alone' (Ambler-Edwards et al. 2009, p. 5), a dollar value is put on the clean air, healthy waterways and intact ecosystems that good farming practices maintain.

However, Lockie points out that PES are not an alternative to good planning and governance, supported by Costanza et al. (2014, p. 152) who state 'global estimates expressed in monetary accounting units … are useful to highlight the magnitude of eco-services, but have no specific decision-making context'. And who pays for the design and compilation of natural capital information? Lockie therefore argues that the effectiveness of PES through a market mechanism will depend on the policy and governance environment in which they operate, that is:

> the capacity of … agencies to define and enforce appropriate property rights, identify and mobilise sellers, solicit trust, act on behalf of absent buyers, monitor PES implementation and outcomes, and reduce information deficits and associated transaction costs … meeting these conditions is not simply a matter of appropriate incentive design but of political decision-making, moral judgment and social learning. Failure to recognise these conditions potentially undermines the effectiveness not only of MBIs but of alternative policy measures taken contemporaneously with MBIs such as community-based natural resource management. (Lockie, 2013, pp. 96–97)

On a more optimistic note, there are also examples of corporations going beyond *compliance* behaviour and creating social knowledge about the pairing of economic growth and environmental governance. Gallagher and Weinthal (2012, p. 144) describe how in 2006 Sir Richard Branson announced that his company, Virgin Group Limited, would invest all profits from its travel companies over the next 10 years in efforts to stop global warming, stating 'we need to make a virtue out of investing in clean technology … and not be ashamed to want to make profits out of it'. This is a very long way from the 1970s, when General Motors were chided in *The New York Times* for pondering the notion of social responsibility (addressing pollution and safety concerns) as *pure and unadulterated socialism.*

Gallagher and Weinthal conclude that while corporate social responsibility (CSR) might have developed as a defensive approach to the regulatory pressure of the state, many businesses now embrace self-regulation and have changed the way

---

[11] Healthy natural capital assets support and enable agricultural production through the production of ecosystem services. Information about the condition and economic value of natural capital of an agricultural enterprise helps land managers to optimise their operations; helps their customers, investors, lenders and others to use capital allocation decisions to incentivise landholders to maintain or invest in natural capital; and helps policy development.

states approach environmental governance (see also Navarro 2006). No doubt, the enhanced ability for sharing individual comment and opinion through the explosion of social media will only see this grow, as evidenced by the success of entities such as *Uber* and *AirBnB*[12], where the ability for both supplier and customer to rate their interaction has enabled robust confidence in the services offered (Yarwood 2016).

4.9 What Is Different Now?

Contemporary Australia is experiencing within political arenas and mainstream media a renewed national focus on the development of Northern Australia, with agriculture promoted as an overt cornerstone of development. While not the first time in Australia's European history that grand visions for the north have captured the nation's imagination[13], this time they are being debated in a globally connected era increasingly conscious of humanity's impacts on, and obligations to, the natural environment – the Anthropocene (Crutzen 2002; Folke et al. 2011; Steffen et al. 2011b; The Stockholm Memorandum 2011; Westley et al. 2011).

Concurrently, and arguably integral to this era, there is at all scales a transition from *government* to *governance*[14] as around the world communities question the 'national interest paradigm in which the inalienable right of the state … is universally recognized … [as having] the capacity to decide actions concerning a national territory' (de Sartre and Taravella 2009, p. 406). The diverse and divergent motivations, perspectives and 'grass-roots' advocacy generating much of the attention on Northern Australia are at odds with Habermas's account of the *normal* course of events, wherein:

> initiative lies not with civil society, nor even with parliaments and legislatures, but with senior members of government and the administrative bureaucracy, such that issues will tend to start in, and be managed from the center [sic], rather than following a spontaneous course originating in the periphery. (2010, pp. 191–193)

According to Habermas, communicative power only becomes political power when it 'affects the beliefs and decisions of authorised members of the political system … who remain the ultimate arbiters of the public good, because only their actions are governed by explicit and binding democratic procedures', and debate and dispute in

---

[12] Respectively, the world's largest providers of transport and accommodation, though neither owns any physical property. Their business models derive from pairing service providers with customers via the Internet. Founded in 2009, by 2014 Uber was ranked 48th most powerful company in America with an estimated worth of $US62.5 billion. Founded in 2007, Airbnb has >1,500,000 listings in 4000 cities and 190 countries.

[13] North Australian development has been compared with Halley's comet: 'they both come around regularly, though each time they vary in intensity'.

[14] 'The shift from the formal and centralised administration and regulation of populations and territories better known as "government" to the decentralised mode of political management known as "governance"' (Argent 2011, p. 184).

the 'weak' publics of the press are 'better suited for the struggle over needs and their interpretation than … for the resolution of conflicts'. But does this still hold true in our era of global citizen connectivity, social networks and 24-h news cycles?

The June 2015 release of *Our North, Our Future: White Paper on Developing Northern Australia* demonstrated that, through the combined influence of global trends and the continuing efforts of Northern residents and industry, the issue of Northern development had been accepted by Habermas's 'authorised members of the political system'. The subsequent question to this acceptance though is how acceptance will manifest in actual development? That, only time will tell. Improved regional infrastructure will be an obvious indicator of material committment. However, the extent to which the aspirations of contemporary advocates for northern development are realised will rest largely on the ability of the five out of every hundred Australians who live in the north to convince the 95 who don't (and who live predominately in southern capital cities) that their experience and perspective is situationally important and should be essential contributors to decisions and achieving a shared vision for Northern Australia.

The next chapter addresses the methods used to investigate the experiences of individuals who live and work in this environment.

References

AIPS. (1954). *Northern Australia – a task for a nation*. Sydney: Angus & Robertson.

Ambler-Edwards, S., Bailey, K., Kiff, A., Lang, T., Lee, R., Marsden, T., … Tibbs, H. (2009). *Food futures: Rethinking UK strategy*. Retrieved from Chatham House, London: http://www.chathamhouse.org/sites/default/files/public/Research/Global%20Trends/r0109foodfutures.pdf

Argent, N. (2011). Trouble in paradise? Governing Australia's multifunctional rural landscapes. *Australian Geographer, 42*(2), 183–205. https://doi.org/10.1080/00049182.2011.572824.

Aslin, H., & Russell, J. (2008). *Social impact of drought: review of literature*. Retrieved from Canberra:

ATNS database. (2004). The Wik Peoples v The State of Queensland & Ors; The Thayorre People v The State of Queensland & Ors [1996] High Court of Australia (23 December 1996). *Agreements, treaties and negotiated settlements database, indigenous studies program*. The University of Melbourne. Retrieved from www.atns.net.au/agreement.asp?EntityID=775

AustralianGovernment. (2014). *Aboutthenaturaldisasterreliefandrecoveryarrangements*. Retrieved from www.em.gov.au/Fundinginitiatives/Naturaldisasterreliefandrecoveryarrangements/Pages/AbouttheNaturalDisasterReliefandRecoveryArrangements.aspx

Bennett, M. M. (1927). *Christison of Lammermoor*. London: Alston Rivers Ltd.

Bottoms, T. (2013). *The conspiracy of silence : Queensland's frontier killing times*. Retrieved from http://jcu.eblib.com.au/patron/FullRecord.aspx?p=1190541

Bourdieu, P. (1986). The forms of capital. In J. Richardson (Ed.), *Handbook of theory and research for the sociology of education* (pp. 241–258). New York: Greenwood.

Brown, K. (2011). Sustainable adaptation: An oxymoron? *Climate and Development, 3*(1), 21–31. https://doi.org/10.3763/cdev.2010.0062.

Cheshire, L., & Lawrence, G. (2005). Re-shaping the state: Global/local networks of association and the governing of agricultural production. In V. Higgins & G. Lawrence (Eds.), *Agricultural governance: Globalisation and the new politics of regulation* (pp. 35–49). London: Routledge.

Coleman, J. S. (1988). Social capital in the creation of human capital. *American Journal of Sociology, 94*, S95–S120.

Costanza, R., de Groot, R., Sutton, P., van der Ploeg, S., Anderson, S. J., Kubiszewski, I., … Turner, R. K. (2014). Changes in the global value of ecosystem services. *Global Environmental Change, 26*(0), 152–158. doi:https://doi.org/10.1016/j.gloenvcha.2014.04.002.

Cottrell, A. (2005). Communities and bushfire hazard in Australia: More questions than answers. *Global Environmental Change Part B: Environmental Hazards, 6*(2), 109–114. https://doi.org/10.1016/j.hazards.2005.10.002.

Crutzen, P. J. (2002). Geology of mankind. *Nature, 415*(6867), 23. https://doi.org/10.1038/415023a.

CSIRO. (2014). State of the Climate 2014. www.csiro.au/State-of-the-Climate-2014

Dale, A. (2014). *Beyond the North–South culture wars: Reconciling Northern Australia's recent past with its future*. Dordrecht: Springer.

Dale, A., Vella, K. J., & Cottrell, A. (2014). Can social resilience inform SA/SIA for adaptive planning for climate change in vulnerable regions? *Journal of Natural Resources Policy Research, 7*(1), 93–104. https://doi.org/10.1080/19390459.2014.963371.

Daly, H. E. (1991). Sustainable growth: An impossibility theorem. *National Geographic Research and Exploration, 7*(3), 259–265.

de Sartre, X. A., & Taravella, R. (2009). National sovereignty vs. sustainable development lessons from the narrative on the internationalization of the Brazilian Amazon. *Political Geography, 28*(7), 406–415. https://doi.org/10.1016/j.polgeo.2009.09.008.

Deloitte Access Economics. (2015). *The economic cost of the social impact of natural disasters*. Retrieved from http://australianbusinessroundtable.com.au/assets/documents/Report%20-%20Social%20costs/Report%20-%20The%20economic%20cost%20of%20the%20social%20impact%20of%20natural%20disasters.pdf

Deloitte Access Economics. (2016). *Building resilient infrastructure*. Retrieved from http://australianbusinessroundtable.com.au/assets/documents/Report%20-%20Building%20Resilient%20Infrastructure/Report%20-%20Building%20resilient%20infrastructure.pdf

Douglass, L. L., Possingham, H. P., Carwardine, J., Klein, C. J., Roxburgh, S. H., Russell-Smith, J., & Wilson, K. A. (2011). The effect of carbon credits on savanna land management and priorities for biodiversity conservation. *PLoS One, 6*(9), e23843. https://doi.org/10.1371/journal.pone.0023843.

Ellyard, P. (2004). *Creating a 21st century sustainable society and built environment in Northern Australia*. Paper presented at the Launch of Centre of Excellence in Tropical Design, Townsville. Retrieved from www.tropicaldesign.org/Launch_PE_Centre_for_Tropical_Design.pdf

Eversole, R., & Martin, J. (2005). *Participation and governance in regional development: Global trends in an Australian context*. Hampshire: Ashgate.

Fache, E., & Moizo, B. (2015). Do burning practices contribute to caring for country? Contemporary uses of fire for conservation purposes in indigenous Australia. *Journal of Ethnobiology, 35*(1), 163–182. https://doi.org/10.2993/0278-0771-35.1.163.

Farmar-Bowers, Q. (2011). Is achieving sustainable agriculture and forestry enough to support country towns? In J. Martin & T. Budge (Eds.), *The sustainability of Australia's country towns: Renewal, renaissance, resilience* (pp. 157–183). Ballarat: VURRN Press.

Flint, C. G., & Luloff, A. E. (2005). Natural resource-based communities, risk, and disaster: An intersection of theories. *Society & Natural Resources, 18*(5), 399–412. https://doi.org/10.1080/08941920590924747.

Folke, C., Jansson, A., Rockstr, J., Olsson, P., Carpenter, S. R., Chapin, F. S., et al. (2011). Reconnecting to the biosphere. *Ambio, 40*(7), 719–738.

Frawley, K. (1999). A 'green' vision: The evolution of Australian environmentalism. In K. Anderson & F. Gale (Eds.), *Cultural geographies* (pp. 265–293). Melbourne: Longman.

Gallagher, D. R., & Weinthal, E. (2012). Business-state relations and the environment: The evolving role of corporate social responsibility. In P. F. Steinberg & S. D. VanDeveer (Eds.), *Comparative environmental politics: Theory, practice and prospects* (pp. 143–170). Cambridge, MA: MIT Press.

Gammage, B. (2011). *The biggest estate on earth: How aborigines made Australia*. Crows Nest: Allen & Unwin.

Gill, F. (2011). Responsible agents: Responsibility and the changing relationship between farmers and the state. *Rural Society, 20*(2), 128–141.

Giorgas, D. (2007). The significance of social capital for rural and regional communities. *Rural Society, 17*(3), 206–214.

Green, D. L. (2006). *Climate change and health: impacts on remote Indigenous communities in northern Australia*. Aspendale: CSIRO Aspendale.

Guha-Sapir, D., D'Aoust, O., Vos, F., & Hoyois, P. (2013a). The frequency and impact of natural disasters. In D. Guha-Sapir, I. Santos, & A. Borde (Eds.), *The economic impacts of natural disasters* (pp. 7–27). New York: Oxford University Press.

Guha-Sapir, D., Santos, I., & Borde, A. (2013b). *The economic impacts of natural disasters*. New York: Oxford University Press.

Habermas, J. (2010). In B. Fultner (Ed.), *Jurgen Habermas: Key concepts*. Durham: Acumen.

Hanna, K. S., Dale, A., & Ling, C. (2009). Social capital and quality of place: Reflections on growth and change in a small town. *Local Environment, 14*(1), 31–44. https://doi.org/10.1080/13549830802522434.

Hawkins, R. L., & Maurer, K. (2010). Bonding, bridging and linking: How social capital operated in New Orleans following hurricane Katrina. *British Journal of Social Work, 40*, 1777–1793. https://doi.org/10.1093/bjsw/bcp087.

Heath, R. (2018). A new climate for drought policy development. *Farm Institute Insights*, *15*(4).

Heathcote, R. L. (1975). *The world's landscapes: Australia*. London: Longman.

Heathcote, R. L. (1994). *Australia* (2nd ed.). New York: Longman Scientific & Technical.

Heathcote, R. L. (2002). Braving the bull of heaven: Drought management strategies, past present and future. *Geography Monograph Series, 7*, 1–38.

Heffernan, O. (2016, March 26). Reviving extinct beasts. *New Scientist*, 8–9.

Holmes, J. (2000). Balancing interests through land tenure reform: Regional contrasts between the Barkly and the Gulf. In R. Dixon (Ed.), *Business as usual? Local conflicts and global challenges in Northern Australia* (pp. 134–154). Darwin: North Australia Research Unit.

Holmes, J. (2012). Cape York Peninsula, Australia: A frontier region undergoing a multifunctional transition with indigenous engagement. *Journal of Rural Studies, 28*(3), 252–265. https://doi.org/10.1016/j.jrurstud.2012.01.004.

Hummels, D., & Klenow, P. J. (2002). *The variety and quality of a nation's trade*. Retrieved from www.nber.org/papers/w8712.pdf

JCU. (2014). *State of the tropics: 2014 report*. Retrieved from Townsville: www.stateofthetropics.org

Jones, C. (2014). Land use planning policies and market forces: Utopian aspirations thwarted? *Land Use Policy, 38*, 573–579. https://doi.org/10.1016/j.landusepol.2014.01.002.

Keogh, M. (2010). Editorial – bushfire policy: Do we need more than just an ounce of prevention? *Farm Policy Journal, 7*(1), iv.

King, D. (2001). Uses and limitations of socioeconomic indicators of community vulnerability to natural hazards: Data and disasters in northern Australia. *Natural Hazards, 24*(2), 147–156. https://doi.org/10.1023/a:1011859507188.

Kueffer, C., Thelen Lässer, K., & Hall, M. (2017). *Applying the environmental humanities: Ten steps for action and implementation*. Retrieved from Bern: https://naturalsciences.ch/service/publications/97610-applying-the-environmental-humanities-ten-steps-for-action-and-implementation

Lankester, A. (2013). *Sustainability on the Australian rangelands: learning, roles in life and sense of place*. (PhD). Retrieved from http://researchonline.jcu.edu.au/29041/1/29041_Lankester_2013_thesis.pdf

Lockie, S. (2004). Social nature: The environmental challenge to mainstream social theory. In R. D. White (Ed.), *Controversies in environmental sociology* (pp. 26–42). Cambridge: Cambridge University Press.

Lockie, S. (2013). Market instruments, ecosystem services, and property rights: Assumptions and conditions for sustained social and ecological benefits. *Land Use Policy, 31*(0), 90–98. https://doi.org/10.1016/j.landusepol.2011.08.010.

Lockie, S. (2014). Personal communication.

Lockie, S., & Measham, T. (2012). Social perspectives on risk and uncertainty: Reconciling the spectacular and the mundane. In T. Measham & S. Lockie (Eds.), *Risk and social theory in environmental management* (pp. 1–13). Collingwood: CSIRO Publishing.

MacKellar, D. (1908). Core of my heart. *The Spectator Archive*, p. 1. Retrieved from http://archive.spectator.co.uk/article/5th-september-1908/17/poetry

Marshall, N. A., Gordon, I. J., & Ash, A. J. (2011). The reluctance of resource-users to adopt seasonal climate forecasts to enhance resilience to climate variability on the rangelands. *Climatic Change, 107*(3–4), 511–529. https://doi.org/10.1007/s10584-010-9962-y.

Maru, Y. T., Stafford Smith, M., Sparrow, A., Pinho, P. F., & Dube, O. P. (2014). A linked vulnerability and resilience framework for adaptation pathways in remote disadvantaged communities. *Global Environmental Change, 28*, 337–350. https://doi.org/10.1016/j.gloenvcha.2013.12.007.

Massey, D. (1995). Places and their pasts. *History Workshop Journal, 39*, 182–192.

Massy, C. (2017). *Call of the reed warbler: A new agriculture, a new earth.* St Lucia: University of Queensland Press.

Mather, A. S. (1993). Protected areas in the periphery: Conservation and controversy in Northern Scotland. *Journal of Rural Studies, 9*(4), 371–384.

Mayere, S., & Donehue, P. A. (2013). Perceptions of land-use uncertainty in Queensland's resource-based regions. *Australian Planner, 51*(3), 212–222. https://doi.org/10.1080/07293682.2013.812673.

McDonald, M. (2013). Discourses of climate security. *Political Geography, 33*, 42–51. https://doi.org/10.1016/j.polgeo.2013.01.002.

Navarro, Z. (2006). In search of a cultural interpretation of power: The contribution of Pierre Bourdieu. *IDS Bulletin, 37*(6), 11–22. https://doi.org/10.1111/j.1759-5436.2006.tb00319.x.

Noble, K. (1997). *The oft-forgotten human dimension of feral animal research and management.* (MSc), Townsville: James Cook University.

Nott, J., & Price, D. (1999). Waterfalls, floods and climate change: Evidence from tropical Australia. *Earth and Planetary Science Letters, 171*(2), 267–276. https://doi.org/10.1016/S0012-821X(99)00152-1.

Oliver, J. (1980). *The disaster potential.* Paper presented at the Response to Disaster Seminar, James Cook University.

O'Meagher, B., Stafford-Smith, M., & White, D. H. (2000). Approaches to integrated drought risk management. In D. White (Ed.), *Drought: A global assessment* (Vol. II, pp. 115–128). London: Routledge.

Onyx, J., Edwards, M., & Bullen, P. (2007). The intersection of social capital and power: An application to rural communities. *Rural Society, 17*(3), 215–230. https://doi.org/10.5172/rsj.351.17.3.215.

Pascoe, B. (2014). *Dark emu. Black seeds: Agriculture or accident?* Broome: Magabala Books.

Pickup, G. (1978). Natural hazards Management in North Australia: Proceedings of and papers arising out of a seminar held by the North Australia research unit of the Australian National University, Darwin, 11–14 September, 1978, (G. Pickup, Ed). Canberra: Australian National University.

Price, O. F. (2015). Potential role of ignition management in reducing unplanned burning in Arnhem Land, Australia. *Austral Ecology, 40*(7), 857–868. https://doi.org/10.1111/aec.12264.

Price, O. F., Russell-Smith, J., & Watt, F. (2012). The influence of prescribed fire on the extent of wildfire in savanna landscapes of western Arnhem Land, Australia. *International Journal of Wildland Fire, 21*(3), 297–305. https://doi.org/10.1071/WF10079.

Price, O. F., Borah, R., & Maier, S. W. (2014). Role of weather and fuel in stopping fire spread in tropical savannas. *Austral Ecology, 39*(2), 135–144. https://doi.org/10.1111/aec.12021.

Putnam, R. D. (2000). *Bowling alone: The collapse and revival of American community*. New York: Simon & Schuster.

Quinn, K. (2016). Contextual social capital: Linking the contexts of social media use to its outcomes. *Information, Communication & Society, 19*(5), 582–600. https://doi.org/10.1080/1369118X.2016.1139613.

Richards, A. E., Cook, G. D., & Lynch, B. T. (2011). Optimal fire regimes for soil carbon storage in tropical Savannas of Northern Australia. *Ecosystems, 14*(3), 503–518.

Richards, A. E., Dathe, J., & Cook, G. D. (2012). Fire interacts with season to influence soil respiration in tropical savannas. *Soil Biology and Biochemistry, 53*, 90–98. https://doi.org/10.1016/j.soilbio.2012.05.009.

Rowell, A. (1996). *Green backlash: Global subversion of the environment movement*. London: Routledge.

Russell-Smith, J., Cook, G. D., Cooke, P. M., Edwards, A. C., Lendrum, M., Meyer, C. P., & Whitehead, P. J. (2013). Managing fire regimes in north Australian savannas: Applying aboriginal approaches to contemporary global problems. *Frontiers in Ecology and the Environment, 11*(s1), e55–e63. https://doi.org/10.1890/120251.

Sarokin, D., & Schulkin, J. (1993). The necessity of environmental economics. *Journal of Environmental Management, 38*, 259–280.

Schuller, T., Baron, S., & Field, J. (2000). Social capital: A review and critique. In S. Baron, J. Field, & T. Schuller (Eds.), *Social capital: Critical perspectives*. Oxford: Oxford University Press.

Schwartz, D. (2018). *Desperate farmers left high and dry as federal relief funding runs out before drought does* [Press release]. Retrieved from http://www.abc.net.au/news/2018-04-28/farmers-high-and-dry-as-federal-relief-funding-runs-out/9704146

Singer, P. (1990). *Animal liberation* (2nd ed.). London: Cape.

Sloyan, M. (2014). The costs and benefits of animal welfare: How the United Kingdom pork industry adapted to changes in animal welfare. *Farm Policy Journal, 11*(1), 33–37.

Stafford-Smith, M. (2005). Living in the Australian environment. In L. Courtenay-Botterill & D. A. Wilhite (Eds.), *From disaster response to risk management: Australia's National Drought Policy* (Vol. 22, pp. 5–14). Dordrecht: Springer.

Steffen, W., Grinevald, J., Crutzen, P., & McNeill, J. (2011a). The Anthropocene: Conceptual and historical perspectives. *Philosophical Transactions of the Royal Society, 369*, 842–867. https://doi.org/10.1098/rsta.2010.0327.

Steffen, W., Persson, A., Deutsch, L., Zalasiewicz, J., Williams, M., Richardson, K., … Svedin, U. (2011b). The Anthropocene: From global change to planetary stewardship. Ambio, 40(7), 739–761.

Stehlik, D. (2005). Managing risk?: Social policy responses in time of drought. In L. Courtenay-Botterill & D. A. Wilhite (Eds.), *From disaster response to risk management* (Vol. 22, pp. 65–83). Netherlands: Springer.

Stehlik, D. (2010). Achieving more socially sustainable communities. *Extension Farming Systems Journal, 6*(1), 92–95.

Stephens, M. (2010). Bushfire, forests and land management policy under a changing climate. *Farm Policy Journal, 7*(1), 11–19.

Stokols, D., Lejano, R. P., & Hipp, J. (2013). Enhancing the resilience of human–environment systems: A social ecological perspective. *Ecology and Society, 18*(1), 7. https://doi.org/10.5751/ES-05301-180107.

Stone, D. (2012). *Policy paradox : The art of political decision making*. New York: W.W. Norton & Co: W.W. Norton & Co.

Szreter, S., & Woolcock, M. (2004). Health by association? Social capital, social theory, and the political economy of public health. *International Journal of Epidemiology, 33*(4), 650–667. https://doi.org/10.1093/ije/dyh013.

Taylor, B. M. (2010). Between argument and coercion: Social coordination in rural environmental governance. *Journal of Rural Studies, 26*(4), 383–393. https://doi.org/10.1016/j.jrurstud.2010.05.002.

The Stockholm Memorandum. (2011). The Stockholm memorandum: Tipping the scales towards sustainability. *Ambio, 40*(7), 781–785.

Vanclay, F. (1997). The social basis of environmental management in agriculture. In S. Lockie & F. Vanclay (Eds.), *Critical landcare*. Wagga Wagga: Centre for Rural Social Research, Charles Sturt University.

Vanclay, F. (2003). The impacts of deregulation and agricultural restructuring for Rural Australia. *The Australian Journal of Social Issues, 38*(1), 81–94.

Vanclay, F., & Lawrence, G. (1995). *The environmental imperative: Eco-social concerns for Australian agriculture*. Brisbane: Watson-Ferguson.

Westley, F., Olsson, P., Folke, C., Homer-Dixon, T., Vredenburg, H., Loorbach, D., et al. (2011). Tipping toward sustainability: Emerging pathways of transformation. *Ambio, 40*(7), 762–780.

Whitehead, P. J., Purdon, P., Russell-Smith, J., Cooke, P. M., & Sutton, S. (2008). The management of climate change through prescribed savanna burning: Emerging contributions of indigenous people in Northern Australia. *Public Administration and Development, 28*(5), 374–385. https://doi.org/10.1002/pad.512.

Wilhite, D., & Pulwarty, R. (2005). Drought and water crises: Lessons learned and the road ahead. In D. A. Wilhite (Ed.), *Drought and water crises: Science, technology, and management issues*. Dordrecht: CRC Press.

Wilman, E. A. (2015). An economic model of aboriginal fire-stick farming. *Australian Journal of Agricultural and Resource Economics, 59*(1), 39–60. https://doi.org/10.1111/1467-8489.12038.

Yarwood, S. (2016). *City 2050*. Retrieved from http://stephenyarwood.com/

Chapter 5
Why and How to Consider the Resilience of Individuals?

That is the beginning of knowledge – the discovery of something we do not understand.
Frank Herbert (1981)

5.1 Introduction

In this chapter the author's ontological and epistemological positions within the interpretative paradigm are described, and there is reflection on how these have influenced and underpinned the methodological approach. The emergence and evolution of this approach within the social sciences is also described, as it is an important element of both how the study was approached and why it is important. The methods used for data collection and analysis are described and the emergent research themes introduced.

5.2 A Problem-Focus, as Distinct from a Topic-Focus

In qualitative research, method is primary, as it assists in protecting findings from a researcher's personal bias and assures objectivity; but the practice of qualitative study requires more than simply applying techniques (Polkinghorne 2006). Qualitative researchers study phenomena and processes in their natural settings to make sense of matters in terms of the meanings that people bring to them, which has been expressed as a naturalistic or an interpretative approach to the world (Hallberg 2006) where researchers 'strive to come further than just pure description … they want to gain an in-depth understanding and explore meanings and processes of everyday life' (Hallberg 2013, p. 1). Marton (1986) describes phenomenography as a qualitative research methodology within the interpretivist paradigm that enables

Keith Noble has contributed more to this chapter.

K. Noble et al., *Agriculture and Resilience in Australia's North*,
https://doi.org/10.1007/978-981-13-8355-7_5

the qualitative investigation of the different ways in which people experience or think about something, while Polkinghorne (2006, p. 72) reminds us that qualitative study is 'about producing solutions to show data can be generated and collected that are rich enough to exhibit the detail and depth of the experience under study, and to show meaning can be drawn from these data through an analysis that intensifies its understanding' – that is, problem-centred.

The objective was, through investigation of the operational environment that both influences and contributes to the ability (or otherwise) of individual North Australian farmers to survive and prosper, to understand and describe their resilience strategies. It was never an intention to establish some external metric of what might constitute success or to rank the relative success of individuals either within or between industries. The adaptive capacity of individuals within their enterprise was a principal focus and the investigation undertaken through phenomenographic analysis[1] within a qualitative research methodology. In adopting this approach, I was conscious of the warning by Darnhofer (2014, p. 476) that while:

> Resilience thinking offers a way to conceptualise uncertainty and dynamics, it raises methodological challenges … [as such thinking requires development of processes that do not] … solely focus on analysing what 'is' but on understanding processes, especially the conditions that enable such processes.

This is why a description of the operational and social context of Northern Australian agriculture was provided in previous chapters, an understanding assisted through the author's career in agriculture and natural resource management, his experience as a farmer and industry advocate and his ongoing involvement in Regional NRM organisations. While not purporting to be unique or any special authority, this experience has shaped a contextual understanding that represents an ontological position while definitely assisting researcher credibility and ability to engage participants, and my epistemological position within the interpretative paradigm is predicated upon the belief that 'a strategy is required that respects the differences between people and the objects of the natural sciences and therefore requires the social scientist to grasp the subjective meaning of social action' (Bryman 2001, pp. 12–13).

5.3 A Review of Theory and Validation of the Methodological Approach

Unlike quantitative research, in a qualitative study, the researcher is often the main instrument for data collection, data analysis and data interpretation, which means researchers must acknowledge and identify their inherent biases (Onwuegbuzie et al. 2008). Bias can also occur through the interaction between researcher and

[1] Phenomenography is a research method adapted for mapping the qualitatively different ways in which people experience, conceptualise, perceive and understand various aspects of, and phenomena in, the world around them (Marton 1986, p. 31).

subject and can come to the fore at any stage of the qualitative research process. Miles and Huberman (1994) differentiate such bias into Bias A and Bias B, where:

1. Bias A is the effects of the researcher on the study participant(s) and prevails when the qualitative researcher disrupts or poses a threat to the existing social or institutional relationships and can lead to informants' implicitly or explicitly boycotting a researcher who is viewed as a spy, critic, nuisance, voyeur or antagonist.
2. Bias B is the effects of the study participant(s) on the researcher and can lead the researcher to 'go native', that is, become a participant as opposed to a peripheral-member researcher who develops desirable emic perspective without participating in those activities central to the person/group/society under study.

Bias B was an obvious and significant risk to objectivity in this work, but Strauss (1987) believed bias could itself represent a rich source of data and coined the phrase *experiential data* to describe it, recommending 'mine your experience, there is potential gold there' (p. 11). The important issue is to be aware of the potential for bias, identify when and how it might occur and develop strategies to balance and address it. Phenomenography (as distinct from phenomenology[2]) allows the researcher to use their own experiences as data for analysis and aims for a collective analysis of individual experiences, which Dahlberg (2013, p. 2) describes as the 'approach of in-between' (insofar as it cannot be reduced to any of the '–isms'[3] that stem from the dualism of an outer and an inner world).

Phenomenographers adopt an empirical orientation and then investigate the experience of others in order to extend existing knowledge through a fresh perspective, which presents the researcher with 'aspects of some particular lived experience, which will startle us into recognizing what should have been obvious' (Ashworth 2010, p. 1). In qualitative research the topic is always contextualised and 'the stronger kind of paper within the phenomenological tradition … is one in which … the author, drawing on the evidence of the research participants, personally wrestles with the experience itself' (p. 2). Ashworth concludes that it is the researcher's responsibility to make sense of the research topic and not rely on authorities.

I wanted to identify social theory that could assist the understanding and organisation of mechanisms that improve the ability of those involved in Northern Australian agriculture to manage change, rather than start with a hypothesis as to what these may be. I was particularly interested in considering theories related to the existing socio-political ideology and their subsequent relationship to explicit and implicit policy paradigms and avoid being researchers that have 'persistently misunderstood decision making, and yet have constantly sought to be of influence' (Rist 1994, p. 546). Rist postulates that research would be more useful if reoriented

[2] Phenomenography will lead to better understanding of the perceptions and experiences of a phenomenon, while phenomenology will lead to a better understanding of the phenomenon itself (Bowden 2000; Marton 1981).

[3] Empirism, psychologism, materialism, idealism, objectivism, subjectivism or constructivism, those that stem from the dualism of an outer and an inner world (Dahlberg 2013, p. 2).

away from 'event decision-making' and towards 'process decision-making', which would 'necessitate looking at research as serving an "enlightenment function" in contrast to an "engineering function"' (pp. 546–47).

My approach has been 'inductive' (as described by Neuman 1997, p.46): to build theory from the ground up, whereby detailed observations move towards more abstract generalisations and ideas. This Grounded Theory is a recognised strategy for qualitative research, not just qualitative analysis (Bertero 2012; Seaman 2008), and avoids descriptive interpretations in favour of abstract conceptualisations by the method of constant comparison, which facilitates the discovery of stable patterns in the data – that is, the emergence of concepts. The central tenet of Grounded Theory is that it allows theory to emerge from the data, rather than a previously formulated hypothesis being tested against data.

The 1967 publication *The Discovery of Grounded Theory; Strategies for Qualitative Research* by Glaser and Strauss represented a breakthrough in qualitative research, as it offered methodological consensus and systematic strategies for qualitative research practice (Bertero 2012) and countered the prevailing opinion that quantitative research provided the one and only approach to scientific inquiry (Hallberg 2006). New theory is conceived as the researcher recognises new ideas and themes emerging from what people have said or from events observed. Hypotheses about the relationships can be tested and constructs formed. This approach is flexible and can give voice to the participants, but application of a systematic methodology avoids it being 'just another story'. Grounded Theory methodology is about research questions, data collection, analysis and generating theory. It is not solely data analysis (Bertero 2012), and Hallberg (2006) states that it is not the form in which theory is presented that makes it a theory; it is the fact that it explains or predicts something.

Grounded Theory has developed and evolved since Glaser and Strauss's (1967) 'discovery' and continues to be modified by the era within which it exists. Hallberg (2006) summarises this into three distinct eras of Grounded Theory progression:

1. Glaser's position in the classic mode, which remains close to traditional positivism with an interactionist perspective – a mode occurring in the 1960s
2. Strauss and Corbin's reformulated mode, which moved into postmodernism with an intention to also render the voice of the informants into the results, and driven by a constructivist view of science – occurring in the 1990s
3. The constructivist mode, represented by Charmaz and part of the interpretive tradition and an approach between positivism and postmodernism – occurring in the 2000s

Charmaz (2006, p. 2) herself describes Grounded Theory as a set of methods that 'consist of systematic, yet flexible guidelines for collecting and analysing qualitative data to construct theories "grounded" in the data themselves'. This is important, as it fully repositions Grounded Theory as a flexible approach and not a strict methodology (Seaman 2008). Charmaz describes the process as deceptively simple: (1) read verbatim transcripts; (2) identify possible themes; (3) compare and contrast themes, identifying structure among them; and (4) build theoretical models, constantly checking them against the data.

In consideration of the process of Grounded Theory renewal, Hallberg (2006) argues that because researchers are not necessarily conscious of how an era is shaping their research practice, they should always reflect upon their own ontological and epistemological standpoints (their assumptions about what reality is and how it can be known) when undertaking research. A clear and stated ontological and epistemological position is also recommended by Mason (1996, p. 47), even for explanatory and unstructured data collection, to enable the researcher to be clear about 'what you might be interested in to be able to judge what to pursue in the interviews'. Grix (2002, p. 179) states that 'it is our ontological and epistemological positions that shape the very questions we ask in the first place, how we pose them and how we set about answering them'. Hallberg's conclusion is that this continual renewal of Grounded Theory makes it 'even more qualified as a useful research approach with capacity to manage the complex and continuously changing social world' (p. 148), which reaffirmed my choice of methodology. It has allowed me to place myself in dialectical relationship with the data as well as providing a theoretical vantage point offering its own methodological guidelines and assumptions.

5.4 The Ontology and Epistemology of This Work

The approach was from an Interpretative Social Science hermeneutics[4] philosophical position, which emphasises a detailed reading of 'text' (conversation, written words and pictures) to discover embedded meaning. While a researcher brings their own subjective experience to the text, the intent was, as described by Neuman (1997), to 'get inside it' through detailed study, contemplate its messages and seek connections among its parts and achieve a 'Postulate of Adequacy'[5] – whether it makes sense to those being studied and if it allows others to understand/enter the reality of those being studied.

Acutely aware of the potential for problems that Layder (2013) warns can arise when a researcher identifies with members of social groups, and particularly when they seek to empower them, I disregarded Layder's first recommendation to ensure that personal prejudices and biases don't intrude by 'steering clear of topics or approaches to research that, by their very nature, make it very difficult to maintain a neutral attitude' (p. 3) but endeavoured to follow Layder's subsequent recommendation if one does proceed that the topics be 'handled skillfully and carefully' in order to avoid the criticism of being politically motivated or bipartisan. My justification for continuing despite this potential for conflict is offering the combination

[4] Hermeneutics was originally the practice of interpreting meaning within biblical text but has expanded to include interpretation of text in search of underlying socio-political meaning (Guest et al. 2012, p. 14).

[5] Described by Smart (1976, p. 100) to determine if an Interpretative Social Science theory is true.

of skills and perspectives of life and work experiences and familiarity with the contemporary but largely unwritten protocols in which Northern Australian agricultural industries and its participants operate. As a researcher, this has assisted me to gain access to and understand the views and interpretations of individual participants and to see things from their perspective, bearing in mind Polkinghorne (2006, p. 76) warning that 'we are only at the beginning stage of the refinement of and sophistication in the subtleties entailed in conducting this kind of inquiry'. I leave it to the reader to judge if I have succeeded.

The intention was to record and analyse data both within and relevant to the sector, combined with interpreting individual responses to past, present and future scenarios. It was important for this to be done within the physical context (environment) in which the sector operates, as 'the ontological distinction between humans and nature is breaking down' (Delanty 2005, p. 5) as a consequence of nature's re-emergence in natural and social science in response to the ecological crisis and the growing need to have 'an ethical engagement with the future' (p. 172).

5.5 Methods and Strategy of Data Collection

Listen more, talk less ... ask real questions. (Seidman 2013, p. 86)

Data collection for this work was through two distinct nonsequential processes, the first literature-based (and mindful of the concerns with interpretation of mute evidence[6], particularly that from another era) and the second in-depth semi-structured interviews augmented by attendance and interaction at industry events, conferences and contemplative reflection on my own professional and farming experience.

Ethical approval for this study was provided by the James Cook University (Australia) Human Research Ethics Committee *Ethics Approval Number H5355.* Interview participants were provided with an informed consent form, which was signed by them prior to the interview commencing.

Social research is emergent, with each interview a unique encounter between distinct social actors. Mitchell and Irvine (2008, p. 34) warn researchers to be aware of this 'research footprint' and their need to be reflexive and responsive to its impact on participant's emotional wellbeing. 'Considering the unpredictability of social research, researchers should think about and plan for how they might respond if and when uncomfortable situations might arise'. When performing qualitative interviews, Tanggaard (2008) recommends researchers pay particular attention to situations where interviewees object to what might be thought, said or written about them, because while objections could be the result of a failure to establish rapport,

[6] 'Written texts and artifacts which, unlike the spoken word, endures physically and thus can be separated across space and time from its author, producer, or user; and so have to be interpreted without the benefit of indigenous commentary' (Hodder 1994, p. 393).

they could also constitute a valuable aspect of the interview and enable better understanding (though the interviewer should not provoke such situations).

Open-ended questions defined topics but provided opportunity for further discussion. This is important, as the quality of the results of a qualitative study depends on the study remaining *problem-centred* rather than strict adherence to a sequence of steps and deliberation on what to do next will require sensitivity in response to a prior action – 'simply mechanically following through on a set of previously determined series of actions would not bring about the intended consequence' (Polkinghorne 2006, p. 73).

With regard to the relationship between researcher and interviewee, Latour (2004) thinks neither distance nor empathy define well-articulated science. To be useful, both should be subservient to the touchstone: do they help maximise the occasion for the phenomenon at hand to raise its own questions against the original intentions of the investigator, including, of course, the empathic intentions? Tanggaard (2008, p. 17) believes it is clear from Latour's formulation that 'abstaining from biases and prejudices is a very poor way of handling a protocol' and supports Latour's belief that the path to science requires a passionately interested scientist who provides his or her object of study with many occasions to show interest and to counter his or her questioning. I was heartened when, about halfway through the interviews, a participant observed that semi-structured interviews were a good and natural process – 'like getting to know your in-laws over the washing-up rather than sitting round the lounge room – maybe because you don't have to have eye contact', which agreed with Seidman's observation that stories are a way of knowing, even though they are

> hard and sometimes draining, I have never lost the feeling that it is a privilege to gather the stories of people through interviewing and to come to understand their experience through their stories. … Use of in-depth interviews alone, when done with skill, can avoid tensions that sometimes arise when a researcher uses multiple methods. Seidman (2013, p. 5)

It was an early intention to include Northern Australian Indigenous groups and communities – those already involved in agriculture and those not, as any agricultural expansion that does not engage with this growing demographic will be limited in both opportunity and social benefit. I eventually decided not to: not because it was unimportant but because it was so important, and adequate consideration of all dimensions could not be provided within the study. However, while Indigenous farmers were neither actively avoided nor precluded, that none were encountered is telling in itself.

Ideally the study would have been conducted across the entirety of tropical Australia, (north of latitude 23.5°S, the Tropic of Capricorn) and include a representative spectrum of established and potential agricultural industries. Practical considerations of geography and resources did impose limits, and 66 interviews were conducted between November 2013 and December 2015. Interview location was of the subject's choosing and included the farm office, shed, and kitchen table, though some interviews were conducted while accompanying the subject on a routine activity (such as a water run, where I doubled as gate-opener).

5.6 The Use of Orienteering Concepts to Organise and Analyse Data

Any property that conveys resilience to a specific farm at a specific point in time might well be irrelevant at a later point in time when both the farm and the context will have changed. Therefore, resilience in such perspectives should be understood as 'emergent' rather than as a fixed asset – not as a 'being' but as a 'becoming' (Davoudi et al. 2012, p. 304). Through this reasoning, Darnhofer (2010) suggests resilience thinking as a conceptual framework that assists with dynamic and holistic thinking in farm management rather than as a formal theory, and she suggests that although farming has not been the primary focus of resilience thinking, there is no reason why it should not be applicable to it

> through framing a farm as co-evolving with its context, resilience thinking sheds new light on socio-ecological dynamics and may allow the identification of factors enhancing farm resilience so as to achieve sustainable development. (Darnhofer et al. 2010, p. 195)

Such a view of resilience helps focus on the persistence of an individual farm and the aspirations of many family farms as a process to ensure farm continuity and intergenerational succession. This can also contribute to an explanation of why farmers do what they do, particularly since 'farmers are less interested in avoiding risk, than in information on how to benefit from variations' (Darnhofer 2014, p. 473). This concept of resilience thinking aligns with Holling's (1973) seminal paper, being 'less about reducing risk from known (or assumed) developments than about accommodating future events in whatever unexpected form they may take' (p. 21) – a process for survival and advancement.

Farming systems are 'probably too complex and variable in time and space for resilience models to provide specific, or even closely predictive, guidance to farmers', whereas 'resilience thinking promotes qualitative approaches that strive to understand the dynamics of farms and enable self-organisation, adaptability and transformability' (Darnhofer et al. 2010, p. 195). In this way resources can be allocated to strategies that reduce the impact of a wide variety of events and identify emergent opportunities. Darnhofer (2014) suggests that research to operationalise resilience thinking is necessary, especially those concerning the farmer as a decision-maker and that farm management would benefit from including insights from sociologists and historians who have studied how farmers have persisted through turbulent and uncertain times.

Guided by Darnhofer's thinking and because disaster research tends to 'focus on the immediate post-disaster experience rather than the long-term recovery path' (Flint and Luloff 2005, p. 402), interviewees were not asked what made them resilient to disaster. The semi-structured interviews explored a range of topics around the individual's situation and life journey and how they coped with adversity. This deliberate approach was to avoid either of the dominant perspectives on the relationship between environment and society: that of *disaster as agent* or *disaster as social vulnerability*, as both can exclude local agency[7] (the capacity to act in the face of a crisis) from analysis due to the emphasis on external environmental forces or social vulnerability (Flint and Luloff 2005). As Lowenthal (2000, p. 256) writes

> Nature and society have been disaggregated in Western converse for the sake of a spurious simplicity and in the interest of a supposed morality. But they are segregated only in our minds, never on the ground. Only by recognizing their essential unity can we come to terms with environmental risk.

The research focus was on the individual participant in Northern Australian agriculture and specifically on the individuals' self-identity as farmer, grazier or other manner they self-identified as being involved in Northern Australian agriculture. In a review of research paradigms from the discipline of community sociology, Veinot and Williams (2012, p. 860) suggest that 'community-level information production, circulation, and technology projects should be understood in an institutional, as well as interpersonal, context', and I treated this self-identification as their *institutional* context – Veinot and Williams point out

> Human communities are territorial; they have a biotic substructure in which individual human beings compete for resources, [and this] directs the energies of competitive individuals. .... By restricting the competition of individuals, society achieves equilibrium in a process of collective adaptation. (p. 850)

– appropriate when the institution referred to is dependent on direct utilisation of natural resources within a regulated environment.

5.7 Approaching the Data

The next chapter explores identified farmer's strategies to survive and, in some cases, prosper in their various farming endeavours. The identification of these strategies was from the principal themes that emerged through Nvivo® coding of participant interviews. The themes relate to and are illustrated by individual actions as described by the interview participants. The four themes have been named:

1. The individual's *situational awareness and consequent opportunity-seeking ability* – better described in the humorous but insightful phrase *the Black Art of Experience*: understanding the operating context and implicitly and explicitly factoring the variables into both day-to-day and long-term decision-making
2. The individual's *capacity to plan* and to then stick to the plan

---

[7] In the social sciences, agency is the capacity of individuals to act independently and to make their own free choices. This ability is affected by the cognitive belief structure which one has formed through one's experiences, and the perceptions held by the society and the individual of the structures and circumstances of the environment one is in, and the position they are born into. Disagreement on the extent of one's agency often causes conflict between parties; for example, between parents and children (Elder 1994; Taylor 1985).

3. The individual's *capacity to adapt*, which is closely entwined with the previous theme because disasters happen suddenly and without warning, so knowing when to change the plan is important
4. Each individual's *social connectedness* as a mechanism to realise change – undoubtedly the most complex and interwoven theme and includes people's connection to country, family, communities, their livestock and their industry, along with their relation to the wider world

References

Ashworth, P. (2010). Editorial. *International Journal of Qualitative Studies on Health and Well-Being, 5*, 5535. https://doi.org/10.3402/qhw.v5i4.5535.

Bertero, C. (2012). Grounded theory methodology: Has it become a movement? *International Journal of Qualitative Studies on Health and Well-Being, 7*, 18571. https://doi.org/10.3402/qhw.v7i0.18571.

Bowden, J. (2000). The nature of phenomenographic research [online]. In J. A. Bowden & E. Walsh (Eds.), *Phenomenography*. Melbourne: RMIT University Press.

Bryman, A. (2001). *Social research methods*. Cambridge, UK: Polity.

Charmaz, K. (2006). *Constructing grounded theory: A practical guide through qualitative analysis*. London: Sage.

Dahlberg, K. (2013). The scientific dichotomy and the question of evidence. *International Journal of Qualitative Studies on Health and Well-Being, 8*, 21846. https://doi.org/10.3402/qhw.v8i0.21846.

Darnhofer, I. (2010). Strategies of family farms to strengthen their resilience. *Environmental Policy and Governance, 20*(4), 212–222. https://doi.org/10.1002/eet.547.

Darnhofer, I. (2014). Resilience and why it matters for farm management. *European Review of Agricultural Economics, 41*(3), 461–484. https://doi.org/10.1093/erae/jbu012.

Darnhofer, I., Fairweather, J., & Moller, H. (2010). Assessing a farm's sustainability: Insights from resilience thinking. *International Journal of Agricultural Sustainability, 8*(3), 186–198. https://doi.org/10.3763/ijas.2010.0480.

Davoudi, S., Shaw, K., Jamila Haider, L., Quinlan, A. E., Peterson, G. D., Wilkinson, C., Fünfgeld, H., McEvoy, D., Porter, L., & Davoudi, S. (2012). Resilience: A bridging concept or a dead end? "Reframing" resilience: Challenges for planning theory and practice interacting traps: Resilience assessment of a pasture management system in northern Afghanistan urban resilience: What does it mean in planning practice? Resilience as a useful concept for climate change adaptation? The politics of resilience for planning: A cautionary note. *Planning Theory & Practice, 13*(2), 299–333.

Delanty, G. (2005). *Social science* (2nd ed.). Glasgow: Open University Press.

Elder, G. H. (1994). Time, human agency, and social change: Perspectives on the life course. *Social Psychology Quarterly, 57*(1), 4–15.

Flint, C. G., & Luloff, A. E. (2005). Natural resource-based communities, risk, and disaster: An intersection of theories. *Society & Natural Resources, 18*(5), 399–412. https://doi.org/10.1080/08941920590924747.

Glaser, B. G., & Strauss, A. L. (1967). *The discovery of grounded theory: Strategies for qualitative research*. Chicago: Aldine Pub. Co.

Grix, J. (2002). Introducing students to the generic terminology of social research. *Politics, 22*(3), 175–186.

Guest, G., MacQueen, K. M., & Namey, E. E. (2012). *Applied thematic analysis*. Thousand Oaks: Sage.

Hallberg, L. (2006). The "core category" of grounded theory: Making constant comparisons. *International Journal of Qualitative Studies on Health and Well-Being, 1*(3), 141–148. https://doi.org/10.1080/17482620600858399.

Hallberg, L. (2013). Quality criteria and generalization of results from qualitative studies. *International Journal of Qualitative Studies on Health and Well-Being, 8*, 20647. https://doi.org/10.3402/qhw.v8i0.20647.

Herbert, F. (1981). *God emperor of dune*. New York: Putnam.

Hodder, I. (1994). The interpretation of documents and material culture. In N. K. Denzin & Y. S. Lincoln (Eds.), *Handbook of qualitative research* (pp. 393–402). Thousand Oaks: Sage.

Holling, C. S. (1973). Resilience and stability of ecological systems. *Annual Review of Ecology and Systematics, 4*(1), 1–23.

Latour, B. (2004). How to talk about the body?: The normative dimension of science studies. *Body & Society, 10*(2/3), 205–229.

Layder, D. (2013). *Doing excellent small-scale research*. London: SAGE.

Lowenthal, D. (2000). A historical perspective on risk. In M. J. Cohen (Ed.), *Risk in the modern age: Social theory, science and environmental decision-making* (pp. 251–257). New York: St. Martin's Press.

Marton, F. (1981). Phenomenography – describing conceptions of the world around us. *Instructional Science, 10*(2), 177–200. https://doi.org/10.1007/BF00132516.

Marton, F. (1986). Phenomenography – a research approach investigating different understandings of reality. *Journal of Thought, 21*, 28–49.

Mason, J. (1996). *Qualitative researching*. London: Sage.

Miles, M. B., & Huberman, A. M. (1994). *Qualitative data analysis: An expanded sourcebook* (2nd ed.). Thousand Oaks: Sage.

Mitchell, W., & Irvine, A. (2008). I'm okay, you're okay?: Reflections on the well-being and ethical requirements of researchers and research participants in conducting qualitative fieldwork interviews. *International Journal of Qualitative Methods, 7*(4), 31–44.

Neuman, W. L. (1997). *Social research methods: Qualitative and quantitative approaches* (3rd ed.). Boston: Allyn and Bacon.

Onwuegbuzie, A. J., Leech, N. L., & Collins, K. M. T. (2008). Interviewing the interpretive researcher: A method for addressing the crises of representation, legitimation, and praxis. *International Journal of Qualitative Methods, 7*(4), 1–17.

Polkinghorne, D. E. (2006). An agenda for the second generation of qualitative studies. *International Journal of Qualitative Studies on Health and Well-Being, 1*, 68–77. https://doi.org/10.1080/17482620500539248.

Rist, R. C. (1994). Influencing the policy process with qualitative research. In N. K. Denzin & Y. S. Lincoln (Eds.), *Handbook of qualitative research* (pp. 545–557). Thousand Oaks: Sage.

Seaman, J. (2008). Adopting a grounded theory approach to cultural-historical research: Conflicting methodologies or complementary methods? *International Journal of Qualitative Methods, 7*(1), 1–17.

Seidman, I. (2013). *Interviewing as qualitative research: A guide for researchers in education and the social sciences*. New York: Teachers College Press.

Smart, B. (1976). *Sociology, phenomenology, and Marxian analysis: A critical discussion of the theory and practice of a science of society*. Boston: Routledge and Kegan Paul.

Strauss, A. (1987). *Qualitative analysis for social sciences*. Cambridge, UK: Cambridge University Press.

Tanggaard, L. (2008). Objections in research interviewing. *International Journal of Qualitative Methods, 7*(3), 15–29.

Taylor, C. (1985). *Human agency and language*. Cambridge, UK: Cambridge University Press.

Veinot, T. C., & Williams, K. (2012). Following the "community" thread from sociology to information behavior and informatics: Uncovering theoretical continuities and research opportunities. *Journal of the American Society for Information Science and Technology, 63*(5), 847–864. https://doi.org/10.1002/asi.21653.

Part II
Lived Experience – Why People, Place, and Services Matter

Introduction

In this section the lived experience of farmers is provided, along with that of practitioners from separate but related fields. Farmer's experiences are interpreted through analysis of semi-structured interviews conducted with them by the lead author, and the findings are classified into four emergent resilience themes in Chap. 6 – the strategies that people use to cope with adversity. This is followed by the consideration by two leading experts of other important external influencers on individuals and their resilience: the place in which they live (Chap. 7), and their access to supporting health systems and services (Chap. 8).

Of course, there are many factors that affect the well-being and success of people in Northern Australia, and their influence and impact will vary from person to person; but in the under-populated and over-extended north the costs of providing *somewhere to call home* along with reliable and accessible health care for all ages and stages of life are generally proportionally higher, and their influence on individuals more profound.

These three chapters also provide an opportunity to consider how policy is currently being delivered across different sectors within the same geographical and cultural landscape, and whether this predominately silo-based approach, often developed within more settled landscapes and communities, is necessarily best for Northern Australia.

Chapter 6
The Resilience Strategies of Individuals

6.1 Theme 1: Situational Awareness and Opportunity Seeking – The *Black Art of Experience*

In any human endeavour, there are individuals who stand out. Whether through financial success, personality, remarkable actions or even notoriety, these are names which come up regularly in conversation. Agriculture is no exception, and considering the relatively small number of people *in the game* in Northern Australia, the same names are repeatedly provided as examples of successful or stand-out farmers. This success often related to a demonstrated ability to understand the operating context, and implicitly and explicitly factor variables into successful decision-making to take advantage of a situation (Fig. 6.1). This *situational awareness* was one of four themes that, along with the *Capacity to Plan*, the *Capacity to Adapt*, and *Social Connectedness* were found to form the core of individual resilience strategies irrespective of industry, scale, or geographic location.

Such an ability was described as 'the Black Art of Experience', humorously illustrating that sometimes, from an outside perspective, such people might appear to be either inordinately blessed with luck or possess arcane abilities bordering on the magical. For example, 'it started raining the day they signed the cheque [for the property] and didn't stop until they sold it'. This is not so different from the ordinary magic described by Masten (2001), whose quest to understand the extraordinary revealed the power of the ordinary as contributing so much to resilience:

> 'Resilience does not come from rare and special qualities, but from the everyday magic of ordinary, normative human resources in the minds, brains, and bodies of children, in their families and relationships, and in their communities'. (Masten, p. 235)

No great intuition was required to appreciate that the common skill possessed by such individuals was an innate understanding of their industry's operating context and an ability to manipulate all the variables in favour of an advantageous

---

Keith Noble has contributed more to this chapter.

K. Noble et al., *Agriculture and Resilience in Australia's North*,
https://doi.org/10.1007/978-981-13-8355-7_6

Fig. 6.1 An existential sign of the times. While dealing with the *here and now* is a part of life, informed decisions improve the likelihood of success. (Photo used with the kind permission of Mark Thomson, Institute of Backyard Studies www.ibys.org)

outcome – situational awareness and consequent opportunity-seeking ability. These were not regarded as rare or unique abilities – it was simply that some people are better than others at turning opportunity into commercial success. A successfully retired grazier explained it thus:

> 'Agriculture's not rocket science, but you do need to combine knowledge with what you do. And you have to stick to something, whether it's bank, sheep or cattle – chopping and changing and chasing things never works. You need morals, but you also need to be a business man and look for the opportunity: when someone is doing it tough is the time to buy, and vice versa when selling. And don't tell your bank manager everything, but never give them a surprise'.

Did a well-developed business mind provide the key to agricultural success? Through NVivo® analysis a word cloud of the 100 most commonly occurring words was generated from the interview transcripts,[1] and the dominant words were 'good', 'work', 'farm' and 'people', embraced by 'live' and 'need' (see Fig. 6.3). Without reading too much into this, it is intriguing these words were centre stage and not business terms like 'investment', 'money' and 'production', which were less frequent and around the periphery.

To be a successful farmer requires more than keeping your eye on the money; it requires a clear understanding of what's being done and why. This should not be

[1]All interview participants provided written informed consent prior to commencement of any interview, and ethical approval for this study was provided by the James Cook University Human Research Ethics Committee, *Ethical Approval Number H5355*.

interpreted as diminishing the importance of financial awareness and robust business practices for good farm management, for as another participant stated, 'knowledge makes management easier'. Vignette 6.1 illustrates how decision-making on farms is shaped by biotic and abiotic influences including economic frameworks, prevailing agro-ecosystems, social norms and weather, and actual decisions are decisively influenced by the individual's perceptions, preferences and risk aversion:

> 'how a farmer perceives and conceptualizes the potentials and limits of his or her farm, the risks emanating from economic, social or ecological changes, and the options that he or she can employ to face them' (Darnhofer et al., 2010, p. 192).

Vignette 6.1: Situational Awareness

As a banana farmer, I thought the disease Tropical Race 4 Fusarium was the single greatest threat to my livelihood. I was wrong. Turns out it was the biosecurity response that I really had to worry about. But we got through it, though it took a lot of grunt to establish that the first test was a false positive. But I realised there was another advantage to my business model of leasing rather than owning farms: I could have just walked away. I couldn't do that if I owned it, and sure as hell no one would want to buy it, so I would have been stuck. Not that I was ever going to – I care too much about my staff and my community to abandon them – I'm just saying I could have.

I decided to farm without owning the farm when my Dad retired. Neither my brother nor I could buy him out, so he sold up and went off to enjoy life. I'd seen the anguish for Dad to arrive at that decision and figured I didn't need it. I might miss out on the capital appreciation, but that's in never-never land, while cash flow is today, and it's worked out alright. Think about it: Woolworths don't even own the shelves in their supermarkets! But I've bought a couple of farms now, because the bank won't lend without security. But they're in another farm sector, one where you wind the handle and the money comes out. One is 1700 km away, but I can run that farm over the Internet, with technical support from Denmark.

I know what I do isn't for everyone, and I understand emotional attachment to the land. But I've proven I can do it in a number of industries now, and I reckon there's opportunity for me to do it bigger, particularly with the foreign investment interest in Northern Australia. It seems there are more people with money in the world than there are people who can run a farm profitably. Which is what really worried me with the biosecurity scare – that's not an association I want to pop up when someone in Hong Kong googles® me. I'll never know who decided **not** to contact me as a result of that publicity.

A cyclone is devastating, but there's a timeline. The issue is in your control, and you can make decisions – you know when you're going to get fruit back on the market and you can arrange your finances to aim for that – an end point. Having the farm quarantined though .... we wrote the phytosanitary protocol in 2 days to get our fruit back on the market, but it took the department 2.5 weeks to approve it! You just don't have control. But you do have the $1.1 million fine if you stuff it up.

So it was a great relief when the detection was declared a false positive. I support biosecurity, but my experience wouldn't exactly encourage other growers to put their hand up. It wasn't just the financial impact, it put huge stress on our family – I've never been through a time like it. Officials have to consider individuals as well as the entire industry, because the industry is made of individuals and they have to be supported.

I can stand in front of a board and I can talk in a paddock, and my new bank manager recognises not everyone can do that. If I lived in Brisbane, I'd struggle to get noticed, but up here there's lots of opportunities. I did a few other things – Dad wouldn't let me come back till I got a qualification – but I wanted to be a farmer. But not a broke farmer. Farming is a way of life, but it's also a business, and we have to move with the times. We install cameras in our sheds as management tools – for workplace health and safety and Quality Assurance. They're better value than the biosecurity inspector we paid to sit in our shed all day: he arrived after everyone else and was always the first to leave.

The future? Well it's not big corporate development. I've worked there too, and they suck up money and pay no dividends. You need skin in the game and an understanding of how things fit, including the environment. And there's a role for government, particularly in infrastructure, and keeping all players honest; but they can do it better than my biosecurity experience. You know the Ord's chia industry is the result of one man's inspiration, but that man couldn't have built the infrastructure that enabled it. There's plenty of demand from Asia but will their price point meet ours? Business costs are high here, and you can't turn primary production on and off.

The biggest risk? It's the growing disconnect between cities and farms: people take food for granted and don't stop to think about how it happens. They pay a dollar a litre for milk but take for granted it's safe to drink; pay $8 for a hot Woolies chook, not realising the farmer only gets 70 cents of that to hatch, house and process it. But farming's exciting, and I love it!

Mark[2] is a relatively young farmer who definitely *thinks outside the box*. He grew up on a farm, went away and chose to come back on his own terms. He knew how to grow bananas but chose his location away from the worst of cyclones yet close to a stable agricultural workforce. He values his workers, and they stick by him, but he uses technology to improve systems. He has farmed bananas, cashews, biofuels, macadamias and chickens in both family and corporate entities and has successfully partnered with others to deliver mutually beneficial outcomes.

But possibly his greatest accomplishment to date was keeping his cool during the Tropical Race 4 Fusarium biosecurity scare. He didn't go public or rant in the press –

[2] Mark gave permission to be identified –'that's part of why I'm talking to you'. This was part of his deliberate strategy to address negative publicity arising from his association with the Tropical Race 4 disease outbreak.

his response was always measured, considered and courteous – but firm. He explored options behind the scenes and negotiated in-confidence with government, his eye always on the bigger picture. He looks for and finds opportunity, but never at the expense of his family or community. Mark values his ethical position above all else.

6.1.1 A Good Way of Life

Most people interviewed were born into farming families (not established pre-interview). Of those not born to a farm, all except one had a family or historic connection to agriculture either directly or through marriage. All interviewees spoke of the lifestyle attractions of farming, but no one said they went farming to get rich.[3] This family association also applied to those working in agriculture-related fields but not directly farming, such as research, extension and industry associations.

When asked why they were farming, all participants indicated it was something they had thought about, and no one indicated farming was a default career they were doing without thinking about it. A majority of those born on farms had left at some point to pursue education, secure a trade or try an alternate career, although for many this had been at their parent's insistence: 'Dad wouldn't let me come back till I'd done a trade'. The decision to return to the farm was voluntary in all but one instance, in which it was a timing issue 'I sometimes wonder what would have happened if I'd spent another year ...' – part of a large successful family operation, this person was not keen to see his children continue farming.

For another, the decision to return required an extended period away before realising 'I thought I needed a bloke to come home and farm, but it wasn't so. Not many people get the opportunity to make important life decisions a second time'. After a successful investment banking career followed by time on the international horse training circuit, she realised that 'I could have anything I wanted in life, but I couldn't have everything' and chose to return to the farm at a time when her parents were looking for assistance:

> 'Mum and dad "sort of" asked me to return ... I do all the cattle buying now. Initially it was confronting when agents wanted to talk to Dad, but we've worked through that. People respect me for what I do. The lifestyle is liberating and there's a great sense of achievement. Farming is a lifestyle, but so is any job: investment banker, Bunnings, sitting at home with your feet on the couch. It's a choice we make'.

Entry to agriculture by nonfarm family individuals was often through an alternate career that facilitated the transition, particularly in emergent industries like tropical fruit where the necessary resources to *get a start* were considerably less than the requisite buy-in to established industries like cattle or sugar cane: 'Tropical fruit wasn't even an industry .... I saw potential as lots of people knew the fruit but couldn't get it'.

---

[3] A few participants did tell me the secret to making a small fortune in farming was to start with a big fortune.

Not everyone was doing well, particularly in the pastoral industries where many were entering their 3rd year of drought and still feeling the effects of the 2011 live cattle export ban. Some were making plans to exit the industry, and some had stopped farming, and not all because they'd retired. But no one said 'I regret farming'. There was a strong indication of desire to maintain industry connection and pride in their association with agriculture.

While there was recognition of the opportunities and constraints of farming, there was unanimous recognition of the lifestyle attractions:

- 'Quality of life is very important, particularly with children until they have to go to boarding school. It was wonderful to have them around'
- 'The kids get to see dad all of the time'
- 'Farming is a way of life but it's also a business. In Brisbane, I'd struggle to be noticed, but up here there's lots of opportunities'
- 'Not good return on capital but … it offers a good way of life, with occasional capital increase'
- 'There's a lot of empathy in the industry for people doing it tough. It's not a callous industry'
- 'Most people in cities are taking in each other's laundry, not creating wealth'
- 'What would make me leave? If my pulse gave out! Really there's nothing I'd rather do. It's more than money, it can't be bought. I want to lead an honest life and leave the place in better condition'.

Even where farms were not doing well and industry exit contemplated, the loss of lifestyle was a major consideration: 'We came back [to the farm inherited from a grandfather] because I didn't want to go through life wondering … It hasn't turned out well, but if we leave there is no way our kids will ever get a start, so we're trying to hang on'.

Stories of hardship and compromise were frequent, though usually not told with regret – they were a part of life to be accommodated. For one interview, prearranged, I arrived to find the participant had been delayed. He arrived 5 h later with 12 decks of weaners which had to be unloaded in the dark, followed by dinner for the road train drivers, but there were no complaints, and everyone enjoyed the evening. It was a part of life. Another said:

> I work 6 to 5 summer and 7 to 5 winter with half-hour smoko[4] and half-hour lunch, 3 hours Saturday, 3 hours Sunday. A weekend off every 6 weeks and Townsville for the V8 supercars. But at 5 pm I sit down and enjoy the view and the birds and I don't ponder *what if*?

There were plenty of grumbles about the lack of services considered basic elsewhere in Australia, and real concerns about the continuing population decline in more remote areas, particularly with the limited or non-existent services that many now feel threatens their ability to maintain functional social systems:

- 'Internet is hopeless. Providers are only interested in signing you up, and that's where the subsidy stops along with the service. That's why so many properties have 4 satellite dishes on the roof. You come in to send an email, then spend 3 days sorting it instead of sorting cattle';

[4] Work break.

- 'The vastness of this country is not going away. Some like the isolation, some the animals, ... but when you lose too much population the social basis is severely diminished'
- 'There used to be a cook and governess to talk to, now there's no one'
- 'It's not the industry that stops bank managers wanting to live here – it's the lack of social environment for wives and families. Twenty years ago, there were tennis parties every weekend ... we've got the only functioning tennis court left in the district'
- 'There was no phone reception and no one on the 2-way [radio] so ... had to drive to the house, then another 2 hours till the ambulance arrived'.

Overall though, the attitude towards such constraints was well summarised by 'We allow ourselves 1 day to have a really good bitch, and then we get on with life. That sort of stuff can really consume you if you let it'. So, lifestyle attraction is important to farmers despite the constraints, but does it contribute to their ability to succeed, particularly when 'the emotional attachment to land makes people stick, but also makes them closed-minded to other possibilities and fearful of leaving'?

Perhaps the answer is in the experience of a successful new entrant to farming: 'My wife was from a farm and we thought it'd be a good place to bring up our kids. I saw that it was banana farmers buying all the new fishing boats, so we bought a banana farm'. This person had been successful in a pre-farm business (providing the ability to buy a farm), thought about what type of farming to enter and then was a successful farmer. His business acumen worked to whatever industry he applied himself, but his decision to farm in preference to other career options was a lifestyle choice.

Some participants had done well financially in the past; some were doing well in the present. Some were about to commit to a major acquisition to provide intergenerational opportunity for their children, while others were the last witness to once vast family empires. Everyone was trying to stay ahead of the bank, but for everyone, the farm meant more than a way to pay the bills – it was part of who they were, and they were proud of it. They farmed because they wanted to farm.

6.1.2 Travel Broadens the Mind

Whether an apprenticeship, a degree, a year driving tourist safari, or time as a jackaroo[5] on other properties, most interview participants had at some stage left the farm and experienced another life. All felt they benefited from the experience. For some, it was the strong desire to return knowing that they had a passion for it, rather than not knowing any different; for others, it provided a fallback income stream for hard times; and for most, it provided skills, experience and knowledge about the bigger world that they brought home and used. Most had been encouraged by their parents

[5] A jackaroo is a young man (feminine equivalent jillaroo) working on a sheep or cattle station to gain practical experience in the skills needed to become an owner, overseer or manager.

to go, rather than return: 'my boys are interested [in the farm] but I won't push them ... they're all getting trades – it's up to them'.

Many participants continue to travel extensively, though not in an organised package-holiday manner: time away from the farm was often related to farm business, training, an industry association or regional NRM body or social events related to their industry or district. Invariably the feeling was what they brought back to the farm was more important than any relief from being away:

- 'After travelling round Australia, I realised what an amazing place home was … my overseas trips teach me more than all the DPI research'
- 'I had four career changes before settling on what I wanted to do. I was the 5th of 6 kids so didn't think there was a future for me on the land, but the passion was there'
- 'I copped flak [from my wife] for being away so much when the kids were little, but it's a sacrifice all fathers have to make. I got more out of it than I put in, it's been a major part of my life, and it's made me money'
- 'Joining the board opened my eyes … I didn't know what I didn't know, especially people management skills'.

This extent of off-farm experience surprised me: 30 years previous, while working extensively across rural Australia, I had met many people who had rarely left either the farm or their district, whereas no participants were now in this category. The intervening period has seen widespread adoption of neo-liberal policies by state and federal governments and commensurate service reduction by banks and other agencies. Population s*hakeout* has occurred as a consequence, leaving those who want to stay and are capable of doing so, while many others have moved through choice or necessity.

An unfortunate consequence of these policies though, combined with hard economics, seasonal conditions and changed industry practices (sheep to cattle, use of helicopter mustering, etc.), has been a significant reduction in regional populations, with subsequent impact on social opportunity and livability. This has been particularly so in the extensive grazing industries and more remote regions, where lifestyle amenity agriculture opportunities are less obvious. Farms have survived, and production maintained, but rural communities have suffered. At what point does the system collapse? 'Do we really want FIFO[6] farms for Northern Australia?'

6.1.3 Off-Farm Influences

Situational awareness is more than understanding the workings of one's own farm. Interview participants demonstrated awareness of a range of external influences and drivers including international markets, human-induced climate change and political trends. No one interviewed saw the world stopping at their farm gate. The perceived ability to respond to such issues varied though, from 'we can't control what happens on the other side of the barbed wire, so we don't worry about it' through to 'I give back wherever I can … [I am supporting] establishment of a Nuffield

[6] Fly-in, fly-out – the predominant employment model for regional mining operations.

scholarship in India'. Being aware of a situation and using this awareness to advantage are not the same thing.

Reflecting on the interviews, participants who demonstrated the latter had more obvious business success. This did not always require active involvement in off-farm activities (though many were); rather it was that these participants thought about and researched historical trends and external influences and incorporated this information into their decision-making – they pondered 'what would I do if … happened'. Such farmers had well-developed scenarios flowing from their 'what if …' questions which will be now discussed, illustrating successful individuals tend to be more *situationally aware* and actively *sought opportunities* than other farmers.

6.1.3.1 Finance

While the importance of lifestyle has been recognised, this did not detract from participants' acute awareness of their financial position. Several graziers talked about how their farm was in fact two businesses: an investment in real estate capable of delivering long-term capital return, supported by a cash flow business such as cattle, hay or even off-farm work. These same producers also described how, in the lead up to the GFC,[7] a rapid escalation of property prices had unsettled the status quo which, combined with a decrease in bank due diligence, resulted in some producers being 'caught out' in the post-GFC crash.

While this has translated into public media concern over bank lending practices and a 2018 Royal Commission, and one participant had lost their property as a direct consequence, overall, the mood was one of caveat emptor: 'it's your responsibility to keep an eye on it [the property market] … people are affected by greed or the worry they'll miss out, but that cycle seems to happen every 40–50 years'. This recognition of the importance of knowing your business was widespread:

- 'it's too easy for a man in the bush to turn 50 and never have done the books, then mum or dad leaves …'
- 'if a farmer lacks business knowledge it limits their ability to understand and grow their business ... they don't pay themselves a proper wage, so don't understand the true costs of production'
- 'our kids are volatile, so we need a financial vehicle that works 30 years into the future if we're to buy more property *[for the kids]* without jeopardising our retirement'
- 'people get interest-only loans and forget that the principle isn't theirs'
- 'if you've got four you can comfortably borrow one. If you've got three, you can carefully borrow one. But many had one and borrowed one'
- 'we're price takers on the world market, so we have to understand it'
- 'it's not a fish & chip shop … you can't just consider the balance sheet'

---

[7] The financial crisis of 2007–2008, also known as the global financial crisis and the 2008 financial crisis, is considered by many economists to have been the worst financial crisis since the Great Depression of the 1930s.

- 'you don't have to own the farm to farm - Woolies don't own their shelves - but at some point you need land to borrow, because banks won't lend on cash flow'
- 'Agriculture still operates in the old paradigm where people with the capital aren't using it, and those with the ideas can't get capital'.

6.1.3.2 Infrastructure

There was a clear belief that better infrastructure makes a place more attractive to new residents, and new residents strengthen the case for better infrastructure, but no one wanted 'temples, and infrastructure shouldn't be ego-driven … facilities have to be carefully considered, and designed for the region', and 'not get lost in ambition'. Improving communications and connectivity was rated as important, though a majority believed that better roads and transport infrastructure would deliver them the greater individual benefit, although 'a new irrigation system would benefit all society'.

No one thought that massive infrastructure investment was about to happen as a consequence of the contemporary focus on Northern Australia, 'infrastructure creates jobs and opportunity, but it's treated like a welfare program rather than strategic return-on-investment … taken control away from local people and decisions made by bean counters in Brisbane' was an oft-repeated sentiment, as were examples given of poor strategic decisions:

- 'Birdsville is going to get cable internet because of the pulling power of the races – it's not a social justice issue'
- 'the mining boom started 2003 and was pretty well over by 2013. Pity more wealth or infrastructure wasn't captured [by regional communities] along the way'
- 'Burdekin stage II was supposed to be built straight away, but all those skills and experience is long gone'.

Perhaps, therefore, infrastructure provision was not seen as the sole responsibility of government and was secured by various means at property and community levels, 'internet is hopeless, so we put up a 60 ft. tower and use our mobiles'; 'I know we're over-capitalised but it's to minimise labour, maintenance and risk', though as a consequence of drought and the live cattle export ban 'many places don't have the resources to do even basic maintenance'. A pastoralist and mayor, who described his philosophy as 'you need a relationship with the people you deal with', described how across regional Australia:

> 'our infrastructure is getting worse as a consequence of local government not having the experience to gauge a reasonable estimate, instead leaving it to big city engineers who get paid a commission, so the bigger the better ... the job goes to a big contractor who can come into a broke community with 30 men and their own gear, then leave with nothing to show for it from the community.
>
> Roads are big business for our shire. Local members should follow funding through and ensure it is delivered by local contractors, so long as they have the skills. Give 10% premium to locals. Many councils don't do this as they're worried about risk, but they don't get value for money or improve their local community'.

His shire had established a pre-qualification umbrella under which local contractors could tender for large jobs, and the community had built a community and recreation amenity lake at the town's entrance for 4% the cost quoted to a neighbouring community for a similar facility. The lake provides visual amenity and community recreation but was built, so when the new doctor arrives in town, his wife's first impression is 'I can live here'.

This importance of local government in regional infrastructure provision was also recognised by two other (current and former) mayors who were also farmers: 'local government can keep infrastructure working well … we're probably in a better position to borrow than the federal government … but it's not our role to fund regional expansion'; 'the people to do the work are on the ground, but getting government agreement ... [can be difficult]'. It was acknowledged that smaller shires were better positioned for a 'hands-on' approach, with more flexibility than larger regional councils.

While there were examples of the capacity of groups to effect change 'Community can do these things, don't always have to go to local Government', describing infrastructure built by the Rotary Club, most thought 'government could make it more attractive for people to live here through help with boarding schools etc.' and that better infrastructure was required to attract people to regional areas, especially schools and hospitals, and 'we should be encouraging immigrants from rural areas rather than wealthy people from cities' along with improved recognition and rewarding the economic contribution made by regional Australians – 'people who create wealth for the nation need to be looked after, and encouraged to keep living there'.

6.1.3.3 Global Drivers

No one believed Northern Australia was about to become Asia's food bowl, though many saw opportunities in Asia's growing middle class, particularly for beef production. Bulk commodities such as rice, cotton, cereals and pulses were considered options in some areas but would require collaborative improvement of transport infrastructure to proceed, and 'a new port at Karumba will require a lot of other services like cool chain, handling and AQIS[8]'. High-value and niche products had opportunity for growth and export but entailed risk and were currently limited to regions with good infrastructure, 'in horticulture you need water today, not tomorrow', and was better suited to smaller growers with more flexibility. Distance to southern Australian markets might be further than some Asian markets, but they don't involve AQIS, international transactions or the risk of political vetoes, so entailed significantly lower risk.

The recent political commitment to northern development was welcomed, but none saw it as guaranteed on-ground activity. Roads have been steadily improving, though there's always more that can be done with *large areas still cut off in the wet.* Road transport is the only real option – 'we used to be all rail but now it's only interested in mines. Freight trains don't even stop here now' and sea freight is 'impossible! Another language'. Internet communication was a wonderful business and social tool, though delivery is piecemeal.

Changes to backpacker visas, whereby a visa extension is available to those who work in regional areas, have made it easier to source labour, and 'backpackers are

[8] Australian Quarantine Inspection Service.

fantastic'; 'you can usually rely on a back-packer working', but does little to deliver regional population growth and skills retention – 'everyone has to be inducted, and then they leave'.

6.1.3.4 Natural Resource Management and Public Perceptions of Agriculture

Awareness of stewardship obligations towards the land and the need for more sustainable natural resource management were widespread, though not often expressed in such terms. There was often recognition of past mistakes or those of earlier generations, although many also spoke of the good management ethos they had learnt from their forebears:

- 'Landcare in this region really took off when people were stunned by how the country **didn't** respond after the 1980's drought'
- 'If my grandfather could see what we've done here, he just wouldn't be able to believe it'
- 'Probably can't take too much more red tape out of industry. Changes under LNP *[government]* were good but we're learning to grow with regulation and if take too much out, the pendulum can swing back. Have to have enough there to protect all sectors – individuals, local government, environment, etc.'
- 'Farmers don't feel compelled to headline their sustainability; they just sort of get on with it. It's a buzz word in corporate and commercial spheres at the moment, but sustainability is normal for us.'

It was not unusual though for tension to become apparent when participants discussed environmental aspects of their farming operations. This tension centred on two factors, the first being the need to make a living while not overexploiting or degrading their natural resource base; the second, and more concerning, was the increasing level of enquiry and control from the broader community into farming practices, particularly when this enquiry emanated from specific issue sectors such as some conservation and animal welfare organisations. It was not that participants thought the public had no right to know about their stewardship practices; rather their concern was about misconceptions and subsequent policy decisions, as graphically demonstrated when the *4 Corners* 2011 story on Indonesian abattoir mistreatment of cattle imported from Australia resulted in immediate cessation of all live cattle export, creating chaos throughout Australia's beef industry, particularly Northern Australia. 'The live trade has always fluctuated, but the export ban fiasco put a huge lump in the snake that has taken some time to work its way through the system'; 'it put 300,000 stock into the domestic system that took a long time to clear. Stock in poor condition just got hammered'.

Apart from the financial impact of this action (compounded by the ensuing drought across much of Northern Australia), graziers and farmers alike were concerned that they were *tarred with a broad brush* with respect to their environmental management practices. Many spoke proudly of their commitment to good environmental management:

- 'you should see the fish I catch from the streams running through the middle of my *[banana]* farm ... people come from all over Australia to fish - that wouldn't happen if I was stuffing things up'
- 'I'd love it for people to come out here and see what a good job we do'
- 'Soil health is the new paradigm'
- 'Would love the city to know we did a wonderful job, because we do'.

Participants did not say their industry was without fault or that environmental regulations were unreasonable, but they were concerned that the standard policy response to bad practice seemed to be the introduction of more onerous legislation, rather than enacting existing rules to prosecute known 'cowboys':

- 'Bureaucracy in any place has a lot to answer for as it often slows things down unnecessarily, but there are bandits in the system who will stuff things for everyone and still get away with it despite regulations. They're the dickheads who deserve to go to gaol'
- 'We're over-regulated and it's unwarranted – why have a blanket approach rather than targeting the known offenders?'
- 'the *[legislative compliance]* bar just gets higher, making it more expensive for everyone except the cowboys, who continue to disregard it and so get a market advantage'
- '*[the government]* is buying the green vote at the expense of reality'
- 'We need understandable and enforced legislation, otherwise it's just spin'
- 'Live cattle export-type shutdown could happen overnight in other industries ... cane has to deal with social media and community perceptions over farming impacts on the Great Barrier Reef'
- 'No one can justify bad practice, but why does a whole industry have to suffer for the sins of a few?'.

The above quotes illustrate a widespread and fundamental frustration recorded on numerous occasions and across different production sectors: while farmers must deal with increasing levels of regulation associated with their legitimate business activities, often their critics (people and organisations who might not agree with aspects of productive land use) 'operate outside the law' with apparent immunity. They appear to be able to use illegal protest mechanisms including trespass and fabrication of stories with impunity. Many participants simply could not reconcile this apparent hypocrisy which often resulted in actual consideration of the issue being lost through frustration about the way it was delivered. However, many were acutely aware of the need to address this issue:

- 'bridging the rural/urban divide is probably more important than on-farm improvements'
- 'I guess the modern challenge is that we have more to deal with than just the landscape and climate effects'
- 'To eat, we either need to grow something or kill something, but that doesn't mean environmental degradation ensues. I don't think we've told that message terribly well'
- 'NRM is moderately mainstream, but very broad church and challenged by the ultragreen. That's appropriate, but still need balance'
- 'I love seeing the health of the landscape improve under my management. I wish the health of the community was improving in the same way'.

Several participants alluded to and some specifically discussed the strategy of 'keeping your head down' or ignoring certain environmental issues, working on the principle that 'by the time the 3-year [political] cycle is over they've forgotten it'. The shortcomings of this approach were acknowledged though:

- 'it becomes a habit, then people let good initiatives go by'
- 'If you got a DMP[9] inspectors would come visiting, so everyone stopped getting them. Then the department cancelled them outright because they said no one was getting them so they weren't needed'.

Concern was expressed at the general lack of processes directed towards improving broader community understanding of farming, along with mechanisms enabling inclusive decision making:

- 'the whole NT community equates to a small town, so why isn't the community better consulted?'
- 'shouldn't be "us & them" [city/country] - if we don't work with them, they will beat us, and we'd all lose'.

While not all interview participants were actively involved with Landcare or regional NRM bodies, those who were involved spoke of the value of such groups, including assisting landholders, achieve better environmental outcomes, delivering funding streams that assist landholders to adopt technologies that improve NRM outcomes and greater good initiatives such as weed, feral animal and erosion management and through providing networks for peer-to-peer information sharing. Participants also spoke of the NRM bodies' capacity to bridge the urban/rural divide and facilitate cross-industry response to emergent issues:

- 'NRM groups are not the enemy …doing a good job, independent arbiters in some ways'
- 'That 10 years *[of national Landcare involvement]* was the most exciting time of my life. We demonstrated real outcomes and bridged the divide between farmers and scientists … there are still people who only mix in their own groups. Some are going broke, some are making money, but running their resource down'
- 'The *[commercial]* fishers realised the … exploration impact on fish numbers – if they *[the fishers]* hadn't been there it would have been too late. Fishers and farmers – they understand what's happening much better than single or special interest groups'
- 'some amazing things happen when we work together'
- 'Some people in conservation are hard case, like *[name removed]*, but she lives here and is part of the community, so we have to work with her because she's committed'.

This final point is important – it illustrates the attitudinal difference towards critics from within versus those from outside a community. There were more acceptance and preparedness to consider the views of someone local, and regional NRM groups provided the 'tent' in which such discussion could be had. An additional point made was that 'Organisations like [NRM body] bring educated people into the community who aren't set in the local paradigms and stigmas.'

Regional NRM bodies were seen as valuable by farmers, in some ways bridging the community capacity void left by retraction of government services and assisting

[9] Damage Mitigation Permit, issued by a government department for limited control of pest native wildlife.

cross-sector collaboration between industry, conservation, local government and the extended community – 'good land management is all about choice and community decision-making. You need sound knowledge to make good choices – science and local knowledge'.

6.1.4 Acts of God, Acts of Parliament, and Other Disasters

Folke et al. (2003) state that change is not a disaster but simply a petty alteration of state, and Holling et al. (1998) point out that what might be a disaster for one system does not necessarily have to be one for another, in the eyes of a farmer a disaster is a disaster irrespective of its origin. All interview participants had experienced natural disasters, from tropical cyclones through flood, fire and extended drought, but many had also suffered because of legislative or policy intervention. There were some who exited the industry as a consequence, though only one directly attributed their exit to a specific disaster event. Even in this case, it was the combination of circumstances at the specific time of the event that determined the impact and subsequent outcome:

> 'I didn't go near the farm for 2 days *[after Cyclone Yasi]* as I knew what it would be like. At 56 years of age I didn't want to start again, so I decided to sell. It was a conscious reasoned decision and I still feel good about it'.

Cyclone Yasi was at the extreme end of the natural disaster spectrum, but a farmer over whom Yasi's eye passed stated 'we don't get droughts here, or fire or floods – just a bit of wind every now and then' – and went on to describe how two major cyclones in 5 years had been a catalyst for change, leading him to trellis his crops. He'd been 'as low as he'd ever been after [Cyclone] Larry, but after Yasi I had direction and knew what to do'. This farmer disagrees with those who think such events are best forgotten: 'we have to remember – same with drought, soil erosion – people forget too quickly, then make the same mistakes', a point reinforced by another farmer, 'as a kid I remember my uncles talking about the 1918 cyclone'.

This does not mean long-term impacts of extreme events like Yasi were not obvious: 'people get jumpy when cyclones are around'; 'we'll never see the industry or the rainforest return to what it was in our lifetimes'; 'insurance has gone through the roof'; 'don't know if I would stay if we had another category 5'; and the impact was particularly severe on the emerging tropical fruit industry, with such an exodus of growers that organised marketing stopped and the industry body collapsed. Here though, an appreciation of the wider situation was important to understand why the impact was so severe:

- This emerging industry was at a point where many growers had just transitioned from part-time farming supported by other incomes into full-time farming.
- The age demographic of growers was mid-1950s to late 1960s.
- The time and resources required to re-establish orchards and then look after them for the 3–5 years until they came into production were beyond many growers.

- Farm sales, which had been occurring at good prices, stopped when prospective buyers realised how vulnerable the industry was.
- People who aspired to this niche farming lifestyle could now work 5 years in FIFO mining and buy a *gold plated one* rather than develop a farm over many years through a parallel career, which had been the industry model till then.

The impact of Yasi on other primary industries (cane, cattle, dairy, small crops) was extreme but nowhere near as catastrophic. As one producer described 'I had 300 cows and 300 fruit trees before Yasi. They all lay down, but only the cows stood up again'. A consequence for the banana industry of two major cyclones within 5 years has been farm amalgamation, with larger growers buying smaller operators, leading to an overall increase in production. In some ways, this just accelerated a process already underway through market forces and 'overall, everyone was better off – big farms employ experienced staff and smaller farmers get to have a regular income, weekends off, and time with their family'.

While there has been considerable public discussion about the influence of climate change on cyclone frequency and intensity, it was not a dominant theme in interviews, either on the coast or further inland. Instead, the pervasive attitude was that farming in Australia has always faced uncertain weather which each industry has developed the skills to manage – *it's been that way since the time of the Pharaohs*, and the improving accuracy of short- to medium-term weather forecasts is improving short- to medium-term planning.

In the extensive grazing sector, the impact of extended drought conditions varied from 'Drought doesn't worry me anymore – make a plan for it and go – manage for worst case scenario' through to 'there's no grass, all our stock are gone, we still owe the bank … we're not sure what to do next', but no one interviewed thought more generous drought relief was the answer. The over-riding sentiment was 'drought is part of the landscape and has to be managed for', and many expressed a sentiment that 'drought policy can be corrupting and prop up bad business – some aren't allowed to fail when maybe it would be better for everyone if they did'.

As ever in the world, disaster creates opportunity – 'in tough times, strong hands receive from weak hands', and there were those who prospered because of better prices for those fortunate to have product, while others had used the down time and NDRRA labour subsidies to improve their packing sheds in preparation for a return to production:

- 'Employment assistance was the best part [of government assistance] as we could keep our staff, clean-up, and get ready for the return to production'
- 'Out of crashes come opportunities – we bought *[another property]* at $4.70 an acre mid-beef crash … cut 1,000,000 *[gidgee fence]* posts at $1 post, and now it's probably worth $200 an acre'.

Exotic disease outbreaks can be disastrous – 'one day you're working, next day you're not', but for one grower, the departmental response was even more onerous:

> 'As a farmer I thought getting [exotic disease] would be the worst thing that could happen to me, but I was wrong. It was the department's response. I would never call Biosecurity Queensland again, and it's the same with the Emerald citrus growers'

A sentiment recognised and acknowledged by other growers – 'Cyclones knock you down, but not out, whereas biosecurity could knock you out'. Growers understood and agreed with the necessity of strong biosecurity measures and rapid response – the cause of their dissatisfaction was the harsh and sometimes inept treatment by government, which could have been avoided through dialogue and more inclusive communication processes. The complexity of this impact was illustrated in Vignette 6.1, which also illustrates the importance of maintaining awareness in such situations, keeping an eye on the *bigger picture*.

An initial impetus for this research was the author's experience with a farming disaster, and this was conveyed to participants prior to interview. Individuals were asked about their own disaster experience as part of the semi-structured interviews, so it would not have been a surprise if disasters had emerged as a major theme, but this was not the case. Disasters were a component of the interviews, but not a dominant one. In fact, participants were generally keen to progress discussion from the disaster event to its consequence, as described in the next section.

6.1.5 Dealing with It

When talking about both past disasters and personal mistakes, the focus of many interviewees was on how they now do things differently. Probably, those for whom past mistakes had been more catastrophic were not available to interview, though some participants were clearly 'teetering on the edge of the abyss'. Awareness of the inevitability of error was widespread, as was the need to learn from it rather than 'beat yourself up ... human error is the error, but adversity is making me a better manager'. This farmer also talked about the need to keep one's own counsel, because 'sometimes the negativity can get you down, like reading the Queensland Country Strife,[10] so sometimes you just have to shut out the noise'.

Many participants talked about the importance of reciprocity and recognised the benefits received from others, particularly where no clear obligation was involved:

- 'Every shed I visit I learn something, but I don't just say thanks and walk away. I present them my point of view, and many say, "great idea, why didn't we think of that"'
- 'how hard is it for you to give back an hour in the afternoon when they've done work for you all day'
- 'I wouldn't sell the recipe *[for my production system]* separate from my farm unless the new owner didn't want to continue growing ... a matter of ethics'
- 'chop-chop[11] is now a dirty word, but earlier it was de facto insurance – *[tobacco]* growers gave to others affected by hail etc., and they paid back when they were able'.

[10] A parody of the popular weekly rural newspaper *The Queensland Country Life*.

[11] Untaxed, unregulated, illegal tobacco.

Fig. 6.2 NVivo Word Cloud illustrating the 100 most often recurring words in interview transcripts, wherein the more often the word occurs, the larger the font

6.1.6 Discussion

While the importance of relationships will be discussed further, the above examples illustrate how actions often relate more to how participants see themselves operating in the world than it does to an anticipated response from others. They demonstrate the operational context of many farmers, and most were keen to progress the interview discussion to their individual learnings and how they used these to make things work better. As examples of participants' strategies and actions to manage risk and uncertainty, they align with four key principles for sustainable adaptation described by Eriksen et al. (2011), being:

1. Recognise the context for vulnerability, including multiple stressors.
2. Acknowledge that different values and interests affect adaptation outcomes.
3. Integrate local knowledge into adaptation processes.
4. Consider potential feedbacks between local and global processes.

The farmers interviewed were confident in their ability to deal with the complexities of their situation, an attitude depicted through humour in Fig. 6.2. The confidence farmers exhibited stemmed from a holistic understanding of their particular situation, knowledge of the resources available to them and having strategies available to utilise these resources. This is in agreement with Wisner et al. (2003) that while a climate-related event may be an external phenomenon, the actual risk is located in society itself, making access to resources and knowing how to use them a critical determinant shaping people's vulnerability, which is 'determined by social systems and power, not by natural forces. It needs to be understood in the context of political and economic systems that operate on national and even international scales' (p.7).

6.2 Theme 2: The Capacity to Plan

> *He, who every morning plans the transactions of the day, and follows that plan, carries a thread that will guide him through a labyrinth of the most busy life.* (Victor Hugo)

Planning is the act of thinking about and then organising the activities required to achieve a goal. A very human activity, and like most human activities, some of us are better at it than others. Financial plans, strategic plans, operational plans, life plans and bad plans – planning is something we do as naturally as breathing. And like breathing, sometimes we do it consciously. Bookcases are full of plans for which the printing of the plan sometimes seemed to have been the end point of the process. Perhaps that is why planning is sometimes viewed by industry as the realm of government and, by association, an excuse to do nothing.

Most people have a natural predisposition to action, particularly during times of crisis, operating on the 'it's better to be doing something than just sit here waiting to be run over' philosophy. The course of a professional career and farming has necessitated the author's attendance and participation on both sides of meetings where officialdom was being urged by landholders to 'do something!', providing opportunity to understand the sense of frustration by landholders with perceived government inaction while appreciating the complexity and precedence implications of official commitment. This experience was often the scenario described when the 'P' word was raised in interviews, and Vignette 6.2 illustrates the attitude towards planning for one participant.

Vignette 6.2: A Girl Needs a Plan

I'm five foot two and fifty kilos, so physically there are things I just can't do – like throw a steer, so I have to be clever. And there're other things I can't do, like drive 130 km into town every time I need a part. You learn to make do and plan ahead. Growing up out here you sort of take that for granted, but other people – sometimes they never get it. I suppose my ex-fiancé fitted that category.

I love horses, and I always leave room at day's end to ride – that's my 'out'. I'd use them for work, but reality is motorbikes are more efficient, and so are helicopters, and probably safer.

It's liberating here – me and my 50,000 acres, and it's amazing how much you can get done in a day. But I'm aware of the risks, living by myself and breaking my leg. Eleven months on crutches – that was a challenge, but we wouldn't have the kitchen mosaic without it. Smashing tiles was therapeutic. I always thought I needed a bloke if I was to come home and farm, so life took a significant u-bend when I decided not to marry. I'm only 27 though, so who knows … I do all the cattle buying now, and there's social interaction there, though some of the agents are still getting used to it.

Fear drives everyone, that's why people are such good sheep, but don't be scared of being scared. My parents said 'don't join the noise' – recognise opportunity, and that doesn't happen if you're used to being told what to do. On the other hand, out here we can all be *Kings in our Grass Castles* – 135,000 farmers in Australia and 135 farm lobby groups – too much independence can be a problem too.

Mum and Dad are chalk and cheese but great mates. Dad's gotten cynical like most old cattle men, and sometimes I think it'd be easier if I were a son who never left home, but our industry needs innovation. We're a strong management team, Mum Dad and me, and I'm pleased that being here has let them go off to do other things, have something else to worry about. Corporates don't give that flexibility or commitment.

We run two businesses: land and cattle. You borrow on land, and it appreciates. You can't borrow on cash flow, but you need it to make the land business work. I enjoyed my time in banking – it taught me how hard it is to make a living, and financial literacy was definitely the skill to bring home and complement Dad's knowledge. I'm numerically dyslexic, but that just made me work harder on the numbers.

But horses are the love, and working those big German and Canadian stables taught me the difference between riding and training horses – you have to be in the moment. I use that every day. I'm terrified of multi-tasking – chasing a cow you can only think 1 s at a time. One hundred percent attention required! And it's meditative, being in the moment.

But plan ahead – make decisions when you still have options, one decision at a time. Not round the kitchen table, because there we sit in the same chairs as when we were kids, and that's how we act. And don't make plans in the paddock – you say what you like in the paddock but leave it there. You know that big decision made years ago? Well sometimes the world turns, and it's not right for anymore, and you just have to get that bit out of your teeth. We go somewhere neutral for decision-making, and we always make time for dream-building.

Fig. 6.3 Love Dream Dance Laugh – a succinct life plan, and smashing tiles was good therapy for a broken leg. (Source: K Noble)

> There're often backpackers here, and they stay in the house. It gives back some of the kindness I received when travelling, and it's mentally good for everyone having enthusiastic young people around. This German bloke though would never listen: New Year's Eve, I told him to fuel the truck and we'd go into town. He came back and asked which was the diesel tank, but not before filling the truck with petrol. So I said 'come inside mate and have a glass of milk'. He asked why, and I told him 'because you're going to learn to siphon'. We were great mates after that.
>
> I made the decision to be involved in this business and in this community, and if a cashed-up buyer came along, I'd probably just buy better country. It's a lifestyle but so is working at Bunnings or banking and playing golf. They all require sacrifice.
>
> I can have anything I want in life, I just can't have everything. The trick is knowing the difference. And you know what? Put that last drop of water on the house lawn, because at the end of a long hot day, coming home to green grass is psychologically invaluable.

Though stereotypically atypical, Jayne demonstrates that through planning, an individual can do whatever they choose to do. Jayne has an older brother, but he's elsewhere in the world, and it's Jayne who has come home to run the family property. The business has been restructured, so Jayne is rewarded for her commitment, but her brother is not excluded should he change his mind in the future. 'People should be given opportunity and rewards but think twice before you lock the gate'. Jayne is operating effectively in a traditionally masculine environment through, rather than focusing on her limitations, thinking of alternate options. Jayne looks for the opportunity in adversity. Jayne makes plans but through continual review knows when to modify them. She learns from prior experience and reflects on these learnings to utilise them in her everyday life. She clearly realises the need to make big decisions and to include relevant parties in the process but also that the little everyday decisions are equally important. Like making a joke out of a mistake as you rectify it, and having an inviting green lawn at day's end, effective planning needs to extend beyond business to deliver a satisfying life (Fig. 6.3).

6.2.1 Plan the Future, Not the Past

The *rules of thumb* (heuristics) farmers use in decision-making are important, for while financial accounting provides valuable insights into a farm's cost structure and profitable activities, they are necessarily backward-looking in a context that can change in unpredictable ways; so the value of 'lessons from the past' in future decision-making is not assured (Darnhofer 2014, p. 473). So what heuristics did the interviews reveal?

Lessons from the past were major determinants in participant's decision-making, but not in a recipe-like manner to reproduce success. Many farmers were avid record keepers, from rainfall and temperature through to production and returns, and quite a few demonstrated a sophisticated knowledge of past trends in property and produce prices and the events that determined the movement. This was particularly the case for larger, established farmers, many of whom were third- or fourth-generation farmers. The over-riding intent was the 'need to keep options open, which depends on making the right decisions along the way', a perspective tempered by the reality that 'you can't plan from the grave'.

The ability to apply past learnings to current situations was demonstrated at various scales, from a 40-year cycle in grazing property prices through to a planting regime to exploit periods of better prices that only became evident after thorough analysis of past seasonal returns. There was also an element of intuitive planning, whereby experience provided a clarity of focus:

> 'after Yasi my feeling was "here we go again". I knew I just had to get started, but I also knew straight away that I wasn't going to replant rambutan'.
>
> 'Being able to deal with life is your ability to develop systems to deal with complex variables of the situation – people get into trouble because they can't think holistically. I hate detail, but I know my limitations'.

6.2.2 The Path of Least Resistance

Successful farm planning was often more centred around what could be relied on to work rather than on maximising output. Participants described plans that made effective use of available resources rather than aspirational strategies reliant on yet to be acquired instruments or technology. This ability to *work with what you've got* and to *make what you've got work for you* was apparent at various scales and across temporal dimensions but often came down to achieving balance between effort and return. This balance is described as *the path of least resistance*, and three key components of the strategy are illustrated here:

1. Conflicting demands on time and resources need to be managed:
   - 'You can't go to town to get a part … make do with what you've got'. Time is the limiting factor, so use it wisely. If you can get by without, or reschedule,

that's what you do. There was also an implication that racing off to town 'at the drop of a hat' could be act of avoidance or procrastination.
- 'The difference between good and bad farmers is basically timing – a half day can make all the difference'. Maintain a strategic perspective, and don't get *lost in the moment* – doing something just for the sake of doing something or because of an arbitrary schedule.
- 'You cut your cloth according to your income' – work with what you've got, and don't overextend yourself.

2. Choose a system or process that can be relied on:

   - 'I'm not sure trellising is the cyclone answer and the pruning would be a lot of labour' – comparing shortcomings of a current management system with the risk of change to a possibly better but unproven system.
   - 'Pencil pushers are dime a dozen, but an ignorant old bugger who knows cattle will always ensure there's cattle *to* sell' – it's what you know, not what you think you know that's important; and demonstrated competence is more reliable than theory.
   - 'We've tried a lot of different things but keep returning to the same old thing – come on cows, need more calves' a comment from an innovative family that had tried many new things, many successful, but they had never forgotten their core business.
   - 'I always waited till a property was overstocked before buying the next one so I didn't have to finance the stock' – a measured rather than an opportunistic approach.
   - 'You can only try three or four times with big [crop] plantings, then you run out of money' – before you start something new, do your research, and consider the risk.

3. When planning new systems or activities, consider ongoing cost and labour:

   - 'Do it once, do it right' – time spent fixing things is not productive, and the uncertainty of not being able to rely on something is an additional source of stress, because 'you can always rely on something to break at the worst possible time'.
   - 'Good partnership agreements are as much about the exit arrangements as they are the purpose for which they're *set* up' – think through the entirety of any process.
   - 'I installed a backup generator after Larry … [expensive, but] it paid for itself after Yasi as we would have been 21 days without water' – the experience and learnings of one disaster were incorporated into planning a new cropping venture.
   - 'I can't afford someone else, so I've set it up this way over 15 years' – operating in an area with limited labour and with a self-recognised high standard, business operation was planned around individual capacity.

- 'I know I'm over-capitalised, but we can run the whole show with two people' – when access to labour is limiting, reducing risk can justify high capital expenditure.

6.2.3 More Than Just Me

Formal planning processes were also evident in succession planning, particularly where accommodating more than one child necessitated additional property purchase. Many spoke of the importance of establishing a financial vehicle capable of navigating unknown futures, such as retirement or family marriage breakups. A component of such transitions was inevitably at what point did the *reins* change hands? Most favoured a gradual transition, even when the incumbents had themselves managed properties from a young age. Such decisions were driven by issues of experience versus enthusiasm but overwhelmingly by the desire to maintain both the asset and the relationship within and between children: 'It must be hard watching yourself operating 20 years ago', speaking to her husband about their son's increasing involvement in the farm. Other telling quotes included 'my old man got a lot smarter after I turned 30'; 'it's the 3rd generation farmer who succeeds … but they succeed on the hard work and experience of their predecessors'.

Concern for, and inclusion of, others outside of direct family in decision-making was also evident in planning, almost as an obligation, though not an arduous one. Examples include the inclusion of staff in recreational and social events, social and professional relationships with neighbours and awareness of how others in the district were travelling – 'when people tell you they're concerned about someone, they could often be facing the same situation themselves'. Possibly, this trait is related to what psychology resilience educators call *nobility*, which reflects the need for self-esteem, self-worth, freedom, order and purpose in life. Nobility reflects a need to give back to society, so part of resilience is the need to be altruistic (Richardson 2002).

Possibly, individuals are valued because of the relatively low population, though there was no evidence of diminished attitude between farms in more settled coastal districts compared with those in remote western regions.

6.2.4 Stick to the Plan, Man

The quote 'you have to stick to something, whether it's a bank, sheep or cattle – chopping and changing and chasing things never works' could be interpreted as a warning against innovation or chosen change, but that was not the intended message. The point being made was not that change was undesirable, but that reactive change without adequate thought or consideration of consequences was unwise – you need to plan things, not simply react to circumstances. This same view was put forward by farmers from many industries, almost all of whom had made deliberate and in some cases extensive changes over time in their farm practices.

No interview participant was 'chasing rainbows' or looking for a 'silver bullet' solution, though many described situations where such an approach had been the undoing of another farm. Often, the situation so described occurred when situations were desperate, implying that such an approach was almost one of *last resort*. Conversely, examples were given in which success had ultimately been achieved after suffering hardship through sticking to a proven methodology 'we cut costs to the bone and did no maintenance till things came good'.

An experienced agricultural consultant talked about 'planful people, who therefore have external markers to refer to ... such as when to start and when to stop' – a plan for a plan. Such a nesting of plans seemed to be a common attribute of many successful farmers. Many of these strategies are illustrated by the preceding Vignette 6.2, as are some of the tools to assist negotiating potential pitfalls of the planning process such as 'say what you like in the paddock, but leave it there' – don't let your temper or your ego lead you into rash decisions, and don't make decisions 'round the kitchen table, because there we sit in the same chairs as when we were kids and that's how we act'; respect the individual and their right to have a different point of view, irrespective of the contextual relationships. The importance of sticking to the plan was evident from the comment that 'you have to play the full 80 minutes to be successful – many games are won or lost at 79 minutes. Decide what you're going to do and stick to it'.

What happens though when an unprecedented event occurs, one for which there is no plan? Natural disasters like cyclones are devastating, 'but there's a timeline. The issue is in your control and you can make decisions ... and arrange your finances to aim for that – an end point'. But socially derived disasters[12] can be unprecedented (the 2011 ban on live cattle export, the immediate quarantining of Far North Queensland banana farms after Tropical Race 4 Fusarium disease was detected). Such events might have been foreseeable, even predicted; but being outside individual farmer's experience meant that they could not forecast the timeline nor adequately manage the factors contributing to the timeline:

> 'we wrote the phytosanitary protocol in two days to get our fruit back on the market [after the farm was quarantined], but it took the department two and a half weeks to approve it! You just don't have control. But you do have the $1.1 million fine if you stuff it up'.

This inability to be able to predict or control timelines was a critical difference between natural and legislative disasters, but the implications for individual resilience are equally applicable to events such as financial recession or global events. In such situations farmers have no metric to decide whether or not to stick to their plan.

[12] A term to describe a disaster that occurs as a consequence of a decision made within society.

6.2.5 *Discussion and Conclusion*

Farmers plan to simplify management, improve productivity and include biodiversity and ecological issues in farm decision-making (Agriculture Victoria 2016), and plans include individual aspirations around profit, lifestyle, family wellbeing, sustainability of production and more. Plans assist farmers through the impact and complexity of the many variables outside their direct control, particularly through recording experiences, their own and others, both in writing and through memory, which in turn provides an opportunity to consider options when making decisions.

While some individuals had more developed/formal plans that they worked to, planning was an integral part of everyday life for virtually all those interviewed, and these plans contributed to their ability to deal with uncertainty and hence to their resilience. Farmers were always wondering 'if I do this, will that happen?' This perspective aligns with the statement by Lorenz (2013, p. 9) that 'Social systems are aware of being within an environment with a given history and with certain expectations of the future and are able to learn and act forward-looking in anticipation of future states'.

It was also apparent some people were better at planning than others, and some at times felt 'lost in their plans', particularly if the plan was perceived as a formula that should not be tampered with. While this attitude was limited, it did appear more evident when farm plans had been developed externally and with limited input from the farmer, where there was sometimes a tendency to see plans as 'set in concrete' or not to have a plan B, whereas more 'planful people' used and adapted plans in response to changing circumstances, which leads us to the next theme – The Capacity to Adapt.

6.3 Theme 3: The Capacity to Adapt

That which does not kill us, makes us stronger[13]. (Friedrich Nietzsche)

Following on from, and closely intertwined with, an individual's capacity to plan is their capacity to adapt the plan, because the nature of disasters is that they happen suddenly and without warning – 'natural disasters always involve some element of surprise' (Kuhlicke 2010, p. 671). Adaptation is a process of deliberate change in anticipation of or in reaction to external stimuli and stress (Nelson et al. 2007) – knowing when to change the plan without falling into a spiral of continual reactive adaptation. Vignette 6.3 illustrates an unusual example of adaption.

[13] Nietzsche's work has been associated with fascism and Nazism, but this appears a consequence of the reworking of his unpublished writings by his sister, who became the curator and editor of his manuscripts after his death, as Nietzsche was explicitly opposed to antisemitism and nationalism (Golomb and Wistrich 2002). Expressions gather their own meanings as they enter the popular lexicon, particularly when used and reinterpreted by successive generations and cultures. This quote was specifically included because it was used repeatedly by several interview participants.

Vignette 6.3: Time for Change

During the 1980–1990s, my Australian Plague Locust Commission (APLC) work entailed extended solo biological survey through remote Australia, regularly visiting the same areas and properties. Central Australia might be geographically large, but as a human community, it's more like an extended village. While the odds of regular unscheduled encounters with the same person might seem slim in such a vast region, I became used to how often they happened. Perhaps because of the sparse population, assisted by the harsh physical environment, friendships developed easily.

The reputation of one fellow (let's call him Mac) preceded him. Mac ran sheep on the very edge of the desert, and on my first transit, I was struck by the number of dingo carcasses strung up at every gate. This wasn't particularly unusual at that time, but the sheer number indicated a very strong commitment to dingo control. I travelled this property many times – the desert/pastoral interface being prime locust breeding country, and it was always the same. Inevitably, 1 day I met Mac.

The APLC was well-regarded by landholders, and 'chasing grasshoppers' generally sufficed as both explanation and validation of one's presence, allowing the conversation to move to more interesting topics such as the weather and the state of the country. But with Mac the conversation quickly turned to dingoes: their numbers, their viciousness, the lack of support and commitment from government and other landholders for coordinated dog control, the decrepit state of the dingo barrier fence and the callow nature of the beast. Mac was evidently a man consumed by dingoes and their destruction and feeling pretty alone in the world on this mission. Adjoining properties, far flung though they were, had long ago switched to cattle, but Mac was adamant that he was going to win this fight. And so it went every time I met him.

One day, a fellow Field Officer returned from survey with an amazing story: he'd met Mac along the road, and there, in the front of his Toyota, was a dingo pup. Alive. With a collar! So next survey, I called by his house, and sure enough, it was unmistakably a dingo, and unmistakably a pet. But the real surprise came when I shook Mac's hand – he'd lost that wild (mad?) eye. I didn't have to ask what had brought about the change – straight up he said, 'if you can't beat them, join them'. He'd switched to cattle, dingo impact on his herd was now manageable, and 'they [dingoes] are quite a nice animal when you take the time to get to know them'.

I never found out what changed Mac: whether it was a slow realisation that he was never going to win that fight or if something in that one pup's eye had caught him? For certain though, Mac was a happier man – awake from a bad dream and enjoying life. People have the capacity to change, to reinvent themselves and to surprise others, and they do it when they're ready.

6.3.1 Pay Attention

Many interview participants stated (and even more inferred) that agriculture was not difficult – 'it's not rocket science' but did require attention to detail and timely decision- making:

- 'It looks easy, but it's not … organising paddocks so you don't run out of grass … dealing with livestock is un-predictable and accidents will happen'
- 'We try to make lots of little profitable decisions'
- 'if there's a $0.30 differential between Townsville and Brisbane I'll send to Brisbane, because it only costs $0.24 extra to get them there'
- '[it's] about being aware and making the right decisions … we talk about it all the time'.

People who demonstrated the attribute of paying close attention to detail considered it essential for their business success, because 'Most people don't know where their business is at or what it's worth'. Application of this attribute was not restricted to property management; it was about *keeping your eyes open* in everyday life – to be aware of opportunities and to avoid pitfalls and well-illustrated by the comment '[driving around the country, I see] so many things that could be fixed easily if only the local member was paying attention'. Paying attention was regarded as an essential life skill – 'We always taught the kids to accept responsibility and blame. By taking risks, they know their capabilities and what the cut-off point was'. While not a tangible asset like fungible assets or natural resources, being able to *pay attention* was clearly an element that contributed to an individual's adaptive capacity.

6.3.2 Maintain Flexibility

Change is a fundamental aspect of any system, and system adaptedness changes as the context changes (Nelson et al. 2007). Maintaining flexibility equates to keeping options open when responding to change, a desirable state actively pursued:

- 'Don't get locked into survival feeding, don't go on agistment – sell down, keep young cows, make sure everything is always saleable. [name supplied] last year had to put beasts into a feedlot to get them to a saleable weight – I would never let that happen'
- 'I always keep a paddock spare'
- 'Try to put yourself in the best position to understand what may happen and counteract unfavourable weather conditions.'

These regimes empowered individuals and allowed them to believe that they were in control of their destiny. While still exposed to danger, they 'also have the ability to transform them into manageable risks' (Kuhlicke 2010, p. 688). Contrast this to the situation in which '[He] sent six decks to Biloela [on agistment] to calve and just settled them in when the place sold and the new owner wanted them off in 2 weeks. Nothing he could do but bring them home to no grass'. The person being discussed was under stress through lack of grass and no local agistment options and stretched himself to get his breeders somewhere safe, only to be put back in the same position minus the freight cost. This did not mean agistment per se was bad, as the respondent had 'learnt through agistment – 10 or 12 properties at any one time over the past 20 years' – that 'it's hard to get certainty with agistment as they're 99% relationship and dealing with people', so when you don't have options, you are vulnerable.

Many described the negative psychological impact that a lack of options generated, which equated with the situations described by Brown and Westaway (2011, p. 325) wherein a country or household 'having low adaptive capacity is likely to have high vulnerability':

> 'As the season got worse and late 2013/early 2014 stock prices plummeted some people had no option but to shoot stock, and this lack of options really affected people – their sense of hope disappeared'
>
> 'When people have sold all their cattle, have no feed and there's no rain forecast – they have no options, and then fear kicks in and mental health issues follow'
>
> 'When people are travelling OK they can wear criticism, but when people feel they're cornered and have no options they react badly ... they come out fighting'.

To avoid this situation, some maintained flexibility through scale 'the alignment of our four properties gives drought protection and economy of scale', some through diversification into other crops or farm enterprises, and in two cases farm tourism, while others achieved it through off-farm investments and alternate businesses including a motel, two local hotels, a (farm) engineering works and an automotive repair shop. In some instances, the diversification was a strategic business decision, while in others the option pursued was through seeing opportunity. In one case it was both: their seasonal labour requirement of 35–40 people was compromised by their remote location, so this family used an historic rural inn license to establish a serviced accommodation venture for young travellers and changed their farm operation to accommodate backpackers new to Australia: 'our farm work is now based on 2 hour shifts, we rotate them in/out of the sun and build up to a full day's work over many weeks'.

Maintaining flexibility was an over-riding mantra – 'if worst comes to worst, we are employable people and we know how to work. We're not frightened by change'.

6.3.3 Go the Extra Mile

No one worked *banker's hours* (a 9–5 day): 'when you work for yourself things change – it's not 9 to 5'. Participants in their seventh and eighth decade described situations where they still sometimes put in long days, whereas younger ones saw this as a natural part of the job, and not a disagreeable one either. Extra effort when required was something that had to be done to secure outcomes or prevent previous effort from being eroded. Sometimes it was sheer excitement – 'I was out there every afternoon [on his recently purchased farm] while still helping the new owner all day [in the business he had just sold]'; sometimes to take advantage of an opportunity, 'I had the experience and now I had the land available'; sometimes through pride 'their farm is always a picture of neatness and organisation ... a great hostess with freshly made cakes, scones, and pikelets … they are very proud of their operation'; sometimes through obligation 'the bonds established with neighbouring properties … have endured for the past century and are typical of the North Queensland cattle industry [when describing collaborative fire-fighting efforts]'; and sometimes because there was no other option 'I do 10 km of fencing a year [by himself] to stay in front'.

A previously described interview told of having to wait for the participant to return home late and unload cattle, and that this generated no stress or anxiety. The atmosphere was very much one of everyone doing what had to be done – a part of the job, a part of life, just what happens. What would have felt out of place would have been if someone was not happy about the situation and made it apparent. The feeling was one of 'a problem shared is a problem halved', and this visibly built and strengthened bonds between people.

6.3.4 Pull Your Belt In

While linked to the previous, the necessity to 'ride out' tight economic times through deliberate austerity had also been experienced by a majority interviewed. Often this was a consequence of disaster, and again those people interviewed were obviously only those who had 'come through' or survived such events:

> '[after the black sigatoka disease outbreak] we had to get back to basics – 35 acres of the best bananas we had, smaller workforce, and gradually built up again … to get debt-free and have control over our life again';
> 'finances were tightly controlled [by the bank] in my younger days … no Christmas hams, so I learnt how to do my own';
> 'after the 1973 crash … bullocks fell from $250 to $50 a head and that went on for 4 years … we did no maintenance till things came good. We were all much younger and reckoned we could ride it out, and we did'.

Many interviewed described this ability as a major difference between family and corporate farms: family farms are 'better able to control costs and hold on in tight times' as they can 'cut their cloth according to their income', though this comment was immediately followed by 'not many get that these days'. This last remark was not directed solely towards farmers, though neither were they excluded from the assessment.

6.3.5 The Past Is a Tool

Multigeneration farms often demonstrated an acute awareness, and understanding of their specific situation was different from but influenced by their forebears. Often this awareness was sometimes accompanied by extensive record-keeping; it was more often the result of ongoing dinner table discussions over many generations:

> 'There have been three different but important phases in our family's history. My grandfather had the establishment phase. He was a visionary sort of guy and introduced 32-volt power to the homestead [and] started developing the place with windmills and wells. Then, my father in the 60s and 70s, the heyday of the beef industry, took it further and did a lot of development in terms of pasture and more water establishment and subdivision and fencing. But I think the challenges for my wife and I, our era from the 80s onwards, is we had more of a community involvement phase and that's where we started seeing more [external] interest in how the land was managed and how that affected us in regulatory impact as well. I guess the modern challenge we have is how to deal more with than just the normal landscape and climate effects that the earlier people had'.

Not all farmers interviewed had thought about how their farm circumstances had changed quite to the above degree, but every person talked about the lessons history could provide. These were far-reaching, from the interaction between seasonal conditions and markets through to the impacts of market regulation, corporate agriculture, generational change and changing climates:

> 'It's in our best interests to manage for the next year and year after. If we take too much out of it this year we won't be able to work the land next year. It's a balance for abundance';
> 'the irrigation was built for tobacco and tobacco was a regulated industry, but that didn't stop its collapse';
> 'people said if tobacco goes Mareeba dies, but in reality, it left a great legacy';
> 'Farming is complicated, especially when setting up from scratch; so big endeavours have big risks … and there are plenty of examples of big ag getting this wrong';
> 'some reach an age where they get comfortable and don't want to change, some run out of energy to change, and some need new "legs" to continue to implement change';
> '[my husband's] family has been here so long that we know it all works out in the end, but sometimes at the expense of one generation for the benefit of others'.

To be of value, the past, like any tool, needed to be used appropriately and with skill:

> 'The problem with using historical patterns to make predictions is we know these patterns are changing – the past is becoming less of a guide to the future, making real-time modelling increasingly important. Seasonal climate forecasts will never be perfect, but they don't have to be'.

6.3.6 Dealing with Failure

The capacity to adapt to failure is included although no interview participants raised it as a subject. Many talked about business ventures they were no longer involved in, including different crops; innovative grazing and management regimes; new marketing systems; and separate stand-alone business ventures either aligned with their farming operation or separate from it. Not once did a participant describe the lack of success as a 'failure', not when they had lost money on the venture and not even where a decision was made to stop farming altogether 'I felt good about the decision to sell. It was a conscious, reasoned decision that included lifestyle opportunities and freeing up finance'. These were regarded as life experience, something to learn from, a consequence of trying new things, even a disaster, but never failure. While people interviewed were still mostly farming, participants did include some who had left of their own choosing and one who had exited as a consequence of a bank foreclosure.

6.3.7 Discussion and Conclusion

Adaptation is a *process,* not a check list of actions that address specific threats, and successful adaptation requires going beyond one-off measures and questioning the assumption that every adaptation will be beneficial (Eriksen et al. 2011). There is extensive literature on the concept of adaptation in response to environmental change in social-ecological systems, particularly with respect to climate change (Adger 2006; NATO Science for Peace and Security Series 2007; Smit and Wandel 2006), and Folke (2006) describes adaptive capacity in such situations as a source of resilience.

Adaptability (or the adaptive capacity) of human systems can be defined as the capacity of 'any human system from the individual to humankind to increase (or at least maintain) the quality of life of its individual members in a given environment or range of environments' (Gallopín 2006, p. 300), particularly when human systems are capable of learning and technological progress. Therefore any fundamental analysis of vulnerability needs to take into account how people construct their own vulnerability (Kuhlicke 2010), as people are capable of preparing for the unknown through combining different types of knowledge, particularly local knowledge which is considered as a '*more adaptive and more appropriate way of dealing with environmental threats and instabilities*' (p.687).

Brown and Westaway (2011, p. 323) state that agency is 'clearly related to adaptive capacity' and aligns with a characteristic Brown (2016, p. 124) describes as resourcefulness – 'the capacities and agency of different social actors and their social ecological system to manage and shape change in both positive and negative ways'. This *capacity to adapt* by interview participants was important at many scales, for as Magis (2010, p. 402) points out:

> 'Members of resilient communities intentionally develop personal and collective capacity that they engage to respond to and influence change, to sustain and renew the community, and to develop new trajectories for the communities' future'.

6.4 Theme 4: Social Connectedness

> *Resilience rests, fundamentally, on relationships.* (Luthar 2006, p. 780)

This was the most complex and interwoven theme to emerge from the interviews. It includes people's connection to family, to communities, to their livestock and industry and to where they live, as well as how they view themselves in relation to the world – no surprise, as human lives are typically embedded in social relationships with kin, co-workers and friends across their life span, and 'no principle of life course study is more central than the notion of interdependent lives' (Elder 1994, p. 6). In their work on the role of networks in transforming Australian agriculture, Dowd et al. (2014, p. 559) state 'individuals with stronger, more informed and more effective networks have been generally regarded as more resilient to generic change events than those with weaker ties', and when these relationships are established and maintained, 'increases in local adaptive capacities materialize, resiliency becomes possible, and community can emerge' (Brennan 2008, p. 59).

This theme is closely aligned with the role of social capital and networks described earlier in supporting adaptive capacity, though Brown and Westaway (2011) point out that increasingly scholars emphasise the importance of an individual's agency in these situations, that is, the capacity of an individual to act independently and to make one's own free choices. This theme has been deconstructed into major components which are illustrated through interview excerpts, though in many cases (as illustrated in the next vignette), there was usually more than one component at play.

Vignette 6.4: Social and Family Connections

My Irish great-grandfather was flooded 2 years in a row in Maryborough, so north he went to the mouth of Mossman Gorge (Far North Queensland). He gave the river flat to his brother because he was over floods, and I've been picking up rocks ever since. That was 1883, and in 1894 he was chair of Mossman's new grower-owned sugar mill. And I was chair when Mackay Sugar bought it in 2012, so I suppose we've been there at the beginning and the end, and now the industry's future is outside individual growers or regions control.

It's a good place to farm, and the mouth of the Gorge was a good spot to own when the ILC[14] built the Mossman Gorge (Indigenous eco-tourism) Centre. Sugar is still important to the town, but with the Port (Douglas) just down the road and Mossman Gorge Centre up and running, there are new opportunities in town. I'm growing cocoa and making chocolate. Everyone likes chocolate – wouldn't that be a good souvenir? TOs [Traditional Owners] not working at the centre could grow cocoa across the river and sell exclusively through Voyager's resorts at Uluru and Mossman. Particularly older people, who don't work in the tourism side. That's what North Australia wants – high-value products that everyone has got a hand in – not cheap bulk commodities for export. We have to be smart.

[14] The Indigenous Land Corporation is a corporate Commonwealth entity, established in 1995, with the purpose of assisting Indigenous people to acquire and manage land to achieve economic, environmental, social and cultural benefits.

> Dad was the first farmer to pay Aboriginals a full wage, before the referendum,[15] and they remember that. Mum's lived next to the community for 50 years and never locked her door. They look out for her. That centre has given real opportunity to people, and they're going to own it.
>
> Pride and ownership are important to everyone, and that's why Reef Regulations did so much damage: you tell farmers they're destroying the place they live in and then expect them to work with you? Come on! It was buying a green vote at the expense of reality, and the big-stick approach killed collaboration. It's hard enough to get a start in farming, and we're not encouraging young people to try. If we are going to open up new country for farming, it has to benefit the people who live there – not some multinational.
>
> It's the same with sugar marketing: Low GI sugar started in Mossman – I got talking to a bloke on a mill tour who turned out to be a food chemist. We must work better together and not get greedy. BSES was the best sugar research body in the world, and the industry stuffed it. Sugar marketing is currently transparent and protected, but no guarantees after 2017. If we lose that, I'll stop growing cane. But it looks like I've finally got the two cane growers associations talking about our marketing future. If we don't, growers will just be slaves to the multinationals, so it's important we all try to prevent that. And getting cane to talk to WWF was a good move – we were never going to get anywhere just blueing[16].
>
> I started growing papaws 20 years ago because a mate needed work, and now I make more out of them than cane. Cocoa is good too. The seedlings are tricky to grow, but there's this girl in NSW who grows them for me. She started out wanting some for natural therapies, and one thing just led to another. She gets all sorts of new plants in now and sends them to me. I'm planting finger and yasou limes. Most new opportunities come through personal networks. There's lots of opportunity, and I do love farming, but I reckon I could live anywhere.

Although a third generation and well-connected sugar cane grower, Don continues to look beyond his established industry position for opportunity. He recognises that opportunities present through the personal contacts he makes but, more importantly, that his ability to successfully pursue them is determined by his existing and possible social connections, for example:

- New horticultural endeavours through a woman who contacted him to source cocoa for natural remedies.
- Starting papaya production to provide work for a friend, but now it earns him more than cane, and he is expanding production.

[15] In 1967 Australians voted overwhelmingly to amend the constitution to include Aboriginal people in the census and allow the Commonwealth to create laws for them.

[16] Fighting.

- Cocoa as a new crop, and the potential for boutique chocolate through his farm's location next to the visitor centre – for both himself and the local community.
- Amicable and fair sale of a portion of his farm for the visitor centre, supported by his long-standing family relations with the local Aboriginal community (Don voluntarily sets aside and maintains an area of his farm containing a traditional burial area that is still used and ensured this parcel was included in the ILC transaction).
- The continuity of the family connection with the town's principal employer – the sugar mill and pursuit of innovative opportunities for it.
- As Chair of an industry peak body, recognition that executive changes within the industry now provide opportunity for factional alignment to better address future challenges and the pursuit thereof.
- There is an honest, approachable, yet unassuming manner evident when meeting Don that encourages knowledge-sharing and collaboration and trust. Don has not used his situation for personal gain; rather there is a sense of 'this is who I am, and because of that, this is what I do'.

6.4.1 Family Relationships

'Be prudent, marry well'. A strong statement, and one many in society might be offended by, particularly if interpreted as either chauvinistic, a business transaction relating to dowry or some other old-fashioned anachronism. None of these were the intent. This advice was provided by the patriarch of a large multigenerational family company, and it was core to his belief of what is essential in his life – family. It was not said flippantly. This was a person of few words, and he chose them carefully and with intent, and the interpretation on reflection of the interview was:

- 'Life on a western Queensland cattle property can be arduous and is not for everyone, so if you intend to stay, make sure your life partner will be prepared to stay with you'.
- 'Both you and your partner have a fundamental and enduring responsibility to the children of your union'.
- 'Family is the most important thing in his life, and he would defend them absolutely'.
- 'A good partnership is <u>the</u> greatest asset to help you navigate life's challenges'.

The family (in all its forms) is the basis of all human societies and social structures (Goldsmith 1978), so not surprisingly almost all interview participants spoke readily, and in many instances passionately, about their immediate and extended family; and post-interview I was often reminded of Montagu's (1942) belief that it is human beings who make a family – not the quantity of them but the quality of them. Descriptions of family were sometimes frank and unembellished but usually displayed genuine appreciation of the relationship – 'you can't put passion in a bottle, but it's easier to maintain it when you have a supportive partner and good health'.

There was widespread awareness of the complexity of family dynamics, particularly the interplay over time and transitioning of responsibilities. For example, the old truism 'My old man got a lot smarter after I turned 30' was closely followed by 'my kids think Dad's getting old, but I'm only now realising how things work'. This was followed by a detailed description of his current internal argument with himself over the need to give youth its head versus the strategic intervention of wisdom and how difficult it was to get this balance right, and particularly the 'need to set up an investment structure so the kids can buy us out, but if they have a bust-up or divorce the asset is protected'.

This issue of succession planning was evidenced in many interviews, and many had engaged a formal process, though this did not guarantee a successful outcome – 'my neighbour spent a lot of money on succession planning but it doesn't seem to have worked. Something's missing'. At the heart of every situation was a genuine desire to 'get it right' for all family members and the asset, though these two aspects were not always in alignment: 'Dad gave me the greatest gift – the ability to go out on my own at 19', and while he would like to afford his son the same opportunity, the 'transition to [his son] is always running into busy [sic] and not formalising'. Partly this related to financial risk when such large amounts of interconnected capital were involved, partly it was because the people currently in charge were still in their prime and enjoying what they were doing, but there was widespread awareness that the 'risk with family ventures when someone wants to exit … can be dramatic and more damaging than shareholders in a corporate situation', and while everyone wanted to prevent this, most could recount instances where things had gone badly.

Sometimes grim situations had delivered good outcomes: industry sources estimated that the two category 5 cyclones in the Tully district had reduced banana grower numbers by 30%, but now 'Many small banana growers work for [large producers] and wish they'd done so years ago. They make a wage now, they get holidays and time with their families, and [the large producers] benefit from a stable and experienced workforce'.

Other industry members discussed the advantages of working for larger corporate entities, such as the ability to work in the industry without having a capital base, the experience gained from moving around between properties and having a secure income. But when it came to the ability to survive, no one thought that the future was going to be one dominated by corporate agriculture, as families and family farms 'provides a really solid base for the area and anyone who comes and buys property next door, they tend to stay'.

The examples in this section reflect the value of both human and social capital in resilience processes: human capital – the development and refinement of individual skills through learning and experience and social capital – particularly bonding social capital, through deliberate adoption of strategies to affect business security and succession while valuing and maintaining relationships.

6.4.2 Business Relationships

The importance of moral capital was evident in most conversations relating to business decisions, particularly as a contributor to the trust required when conducting business over extended distance. Ethical business relationships were deemed important irrespective of the type and scale of farm enterprise, though exactly how they manifested did vary across industries. In horticulture, 'your neighbour is your competitor and that's probably always going to be the case' as producers often sell into an oversupplied market with limited shelf life, and 1 day can make a huge difference to grower returns – 'gold for shit, and shit for gold'. Consequently, many horticulturalists have developed strategies to cope with what most described as a one-sided and opaque relationship with their markets. There were elements of trust required and demonstrated, but often, these were qualified through deliberate moderating mechanisms:

> 'Horticulture marketing is not a fair system. I go to my *[Melbourne]* wholesaler at the beginning and end of every season. It's expensive, but pays off – they know I'm involved. The horticulture code *[of conduct]* doesn't work because farmers will deal with the devil they know. Most of the burn stories are from when someone new comes into the market. Investment would increase if profitability improved, but that would need transparent dealings, and I can't see that happening under the current system'.

There were many examples given of attempts in horticulture to develop more equitable systems through collaborative marketing, but these were often viewed by other growers as 'someone trying to take over the world', and most had foundered through concerns about bridging social capital effects (*better the devil you know than the devil you don't*). It was more common for these concerns to be implied, rather than overtly expressed. There were exceptions, of course, and usually these were where the proponents were able to exert significant market pressure, such as a major banana-growing family who now 'market from all over north Qld, which gives them on-ground knowledge [about product supply]'. This family corporation 'work with the end-user rather than against them … they know what the retailer wants, and deliver it'.

However, this marketing service was made available to other banana growers, most of whom have 'nothing signed on paper, we just choose to do it and they do a really good job'. The family's skill is not confined to marketing – 'they always headhunt good people and get the right person in the right place' – and as a consequence is held in high esteem by both the industry and the local community. The family has invested heavily in building their bridging social capital which, in turn, benefits other growers through access to provision of linking social capital, but both are heavily dependent on their demonstrated strength in moral capital.

Reputation, market support and ethics were not restricted to large farms. In one instance a farmer was 'assisting [name supplied] to develop her propagation business ... It's crucial to the industry future [and] run as a separate business on my farm, so she is on hand to work for me too'. Another grower contemplating the sale of his farm 'wouldn't sell the recipe separately unless the new owner didn't want to continue with [crop], as this would ruin the market for all – a matter of ethics'. This grower had developed a high return farming system that conceivably could be more valuable than his actual property, but while important, money was not the only factor in his decision-making.

In extensive grazing, one family had assisted a couple who previously worked for them to secure equity in another farm they were buying, which the couple then managed through a *flat*[17] business partnership: the skills and commitment of the young couple were retained in the business, and they got a start on the property ownership ladder – a mutually beneficial relationship. I asked why this model was not widespread, and the response was that they had 'never met anyone else we would consider being in partnership with'. Other people are interested in the concept, and both parties have done MLA[18] presentations on the model, but the key was evidently finding the right fit between people, and this was a rare thing.

> A successfully retired grazier explained the balance between business and morals as '*You need morals, but you also need to be a business man and look for the opportunity – when someone is doing it tough is the time to buy, and vice versa when selling*'. Later in this interview a name came up, who was described as '*a hard man*' – the implication being that this person was seen to profit too much from other's adversity, possibly seeking out such situations, and thereby lacking in morals. The implication was that he would not be trusted in business transactions.

Many established industries like sugar and beef often work on a pooled market price, though there were instances of developing specialisation and niche markets here as well through such innovations such as low GI[19] sugar and organic beef. Many spoke of the increasing professionalism 'that farmers don't get credit for … we have to think on our feet and make decisions every day that are worth more than individual [government employee] salaries'; and while established commodities like sugar, wool and eggs had experienced past government intervention to stabilise prices, its cessation was generally seen to be for the better:

- 'It's not up to government to create the customer, but government can be an enabler through market access';
- 'Opening up of the cane industry created opportunities for growth which were previously stifled';
- 'The wool floor price was a good idea @ 400 cents, but got greedy and became a price setter. Wool growers were the puppet and the industry body were milking the system'.

The above comment referred to a quasi-autonomous non-governmental organisation (QANGO) to which the government had devolved power, but such bodies results were viewed very differently by the same farmer when speaking of contemporary industry bodies: 'the 1970's [beef] crash was because we were reliant on

---

[17] *Flat* meant no hierarchy – all partners were deemed equal in decision-making.

[18] Meat & Livestock Australia Ltd. is a producer-owned, not-for-profit organisation that delivers research, development and marketing services to Australia's red meat industry.

[19] Glycaemic Index is a relative ranking of carbohydrate in foods according to how they affect blood glucose.

only the US market. Now, thanks to MLA, we have 20–30 markets and so are less vulnerable. MLA do a fantastic job'.

The importance attached to industry bodies was not consistent, with some held in high regard, while others were deemed necessary, but not of great importance. It did appear that the value attached to such bodies was commensurate with the degree of individual involvement, with some participants actively working to increase their effectiveness. A commonly espoused deficiency of industry associations was in their inability to improve understanding between city and country perspectives, as evidenced by the live cattle export ban though the blame for this decision was placed squarely with the federal government. It appeared that a fundamental role of industry associations was to build bridging social capacity on behalf of industry, but their capacity to do so was variable.

Farmers generally spoke well of industry-focused government staff in regional locations but were often less impressed with performance the more distant the decision-makers were physically located:

> 'the local officers were tremendous, but Biosecurity Queensland can be very brutal on individuals';
>
> 'the Flinders River irrigation idea had a lot of interest, but the moratorium on development on top of the 10 years to develop the plan, plus another 4 years to release water, meant that interested producers in their 20s were now in their 40s, and had either moved on or changed direction'.

Discussions also addressed the importance of loyalty, consistency and good relationships: 'important attributes – stick to agents/banks rather than chop and change. Relationships are important, and they build a regional sense of commitment and community', for example:

> 'Backpacker labour in Tully is working well, but we should provide more leisure facilities for them e.g. basketball courts/movies/a bus to the beach';
>
> '*[Satisfaction]* comes back to people and relationships';
>
> '*Successful individuals can inspire others and the model spreads by imitation, not top-down instruction*';
>
> '*We hadn't much sheep experience but* [mentors] *grew up with them and mentored us through the process*';
>
> 'I've thought about ways of growing my skill set through extending the model – that's partly why I'm talking to you today'.

Some also spoke of sectors they did not trust and why: 'I have a fear and worry of consultants who have failed in their own enterprises, then see a niche market of people who are unsure of their own decisions and milk this uncertainty … I don't use consultants'.

6.4.3 *Relationships with Others*

Population numbers fall quickly as you travel west from Queensland's relatively settled east coast. Obviously, farmers are aware of and in most cases resigned to it but that doesn't mean they don't seek and enjoy the company of others. My impression, developed over many years of work in Central Australia, is that these communities are more extended villages where people know each other and stay informed of individual's movements and activities[20]. And like villages anywhere, they maintain connections to neighbouring villages and beyond.

Another impression from this experience was that when you meet people, you tend to get a more three-dimensional view of who they are. What is meant is that it is harder for people to present and maintain a façade than in larger cities, where human interaction can be more easily restricted to a time and place and avoid the third dimension of knowing about their interactions with others whom you already know. For example, a butcher in Longreach sold tickets at the RSL club Friday nights and ran the pony club on Sundays. Of course, this happens everywhere in the world, but when there are fewer people to talk about, the same names are more likely to occur. As a consequence, people's reputations are open to scrutiny, so embellishments need to be made with care. As another example, an opportunity to interview one prominent member of the grazing industry only came about, through my long-standing friendship with his nephew who had worked on his property many years previous – his endorsement carried weight.

Understanding the importance of these behaviours and relationships is fundamental to understanding the relationships between people in Northern Australian agriculture. It is why the *handshake* banana marketing agreements described works. It is not a fool-proof system, and people still take advantage of others, but to maintain standing and build relationships that can provide support in times of need, what other people think of you is very important. A successful farmer with involvement in cane, cattle, broadacre cropping and mining described his 'fundamental concern with opportunistic people who take advantage of others' and was concerned that in our modern age the 'independent arbiter of God has been replaced by the dollar coin'. At the other spectral end from this philosophical concern was a grazier describing a drought relief parcel she'd received containing 'a package of handmade biscuits with a note and my name written on it' and how this personal connection had touched her and 'demonstrated that people genuinely care'.

From the outside, farming can appear hard and overtly masculine, particularly extensive grazing, but in both prior experience and through these interviews, many examples refuted this. A rangeland researcher described (positively) how the industry had become *feminised* during his 25-year career and how the relationship between pastoralists and conservation managers was now *excellent*. A cane farmer described how 'Dad was the first grower … to pay Aboriginals the full wage, and they remember that'. Despite distance, people seemed able to connect with the right person when needed, such as during a relationship breakup where it was 'good to be able to talk to someone outside the family, but someone who understood the industry'.

Central to all these stories was the importance of relationships, reputation and trust. Trust was built and maintained through all the elements of social capital. It

---

[20] *'While it is true that place and locality are important components, community is far more than a geographic location'* (Brennan 2008, p. 58).

was a strategy for building resilience that people understand and value, and 'new opportunities come mostly through my personal networks'.

6.4.4 Community Relationships

Beyond interpersonal relationships was farmer's relationship with their communities – local, state and national. There was considerable mention of the growing disconnect and lack of understanding between urban communities and farmers, though one person did point out that 'the city/bush divide can be just as real between [local town]/bush as between Canberra/the bush'. The impact sectoral groups could have on farm businesses was of particular concern, and many obviously struggled to understand either the perspective or precisely what it was such groups were hoping to achieve: 'it's one thing when an attack comes from a foreign nation, but when it comes from within' – in reference to the 2011 ban on live cattle export to Indonesia. Many were obviously dismayed that often these sectoral lobby organisations 'seem to operate outside the law with impunity'.

Farmers were acutely aware of their declining political influence and subsequent vulnerability to organised lobbying – their declining linking social capital. For example, 'a live cattle export-type shutdown could happen in other industries overnight, and the cane industry needs to deal with social media and community perceptions'. As a consequence, and aided through the mediation of some regional NRM groups, many primary industries have now moved beyond entrenched opposition to environmental lobby groups, as demonstrated by the Reef Alliance: a QFF, WWF and regional NRM bodies collaborative Great Barrier Reef wide approach intended to advance farmer practices beyond industry BMP and fast track the implementation of innovative practices: a deliberate strategy to build bridging social capital. An industry member reported that 'trust is still variable – at a recent event WWF said without warning that BMP doesn't go far enough', but as a consequence, the Alliance has developed a 'no surprises policy that's working a bit better, though the individual grower relationship with WWF will take a long time to change'.

However, after the Queensland Government's Great Barrier Reef Water Quality Taskforce 2016 report concluded that even if 100% of farmers in the GBR catchment adopted best practice management, it still would not be enough to achieve the established water quality targets; two major integrated projects (MIPs) were funded. These MIPs, one in the wet tropics managed by Terrain NRM and the other in the Burdekin River region and managed by NQ Dry Tropics (both regional NRM bodies), were to integrate and evaluate the combined effectiveness of a range of tools and innovative approaches developed by the farmers and other land managers living and working in the catchments that, once up-scaled, could deliver accelerated progress to the targets and inform ongoing investment across the reef catchments (Great Barrier Reef Water Science Taskforce 2016). Rather than designed top down by scientists, the projects were collaboratively designed by land managers informed by science – a fundamental and paradigm-changing shift in landscape management.

When I asked about the future, there was strong agreement that farmers need the community as much as the community needs farmers: 'things would be a lot worse if we just left each alone – we need each other. We just need to improve interaction and understanding'; and 'we will only get the sympathy vote for so long'. Many growers expressed fundamental agreement with environmental and animal welfare legislation, though many blamed government failures to enforce existing regulations as a reason ever -more onerous rules were introduced – 'we need regulation follow through by government to deal with the cowboys, otherwise it's just spin'. An example demonstrating the value of community was early Darwin society, where the 'Chinese, Japanese, Greeks outnumbered Caucasians, but everyone needed each other', and 'this strong sense of identity continues today without us having actually achieved anything other than all living in a remote city'. This person went on to ponder whether more Australian communities were like this in the past, 'but Darwin has held on to it' as a consequence of its physical isolation.

A number of farmers had recently received on-farm assistance from volunteer-based organisations such as *Blazeaid*, which had its genesis in the devastation of the Black Saturday bushfires of 8 February 2009 and whose slogan is 'Not just rebuilding fences, but rebuilding lives' (see www.blazeaid.com). When these farmers described their experiences, it was apparent that both the recognition of their difficult circumstances and the willingness of complete strangers to assist 'at their own expense' were of equal, if not more value than the actual physical work undertaken – 'they had their own caravans, chainsaws, cooking gear, everything! All I had to supply was somewhere to park their van and a bit of meat … and we had a few good nights swapping yarns'. Some concern was expressed over well-publicised events such as hay runs from southern states to western Queensland which were 'more about the Guinness Book of Records than supplying hay'; along with other mixed messages, 'the Blazeaid-type response is confused – they offer help but advocate against foreign investment, which for some sectors is the future'. These instances reinforced the importance of avoiding competing agendas when delivering services if enduring social capital is to be achieved.

I interviewed one past and two sitting mayors of predominately rural communities (also farmers), and each spoke of the importance of local government in contemporary and future community prosperity. In particular, the importance of maintaining sound relations with the two other tiers of government for provision and maintenance of essential infrastructure was described – 'local government is a business, and the mayor's role is to get money – not fight with them'. The importance of community relations was also emphasised – 'If you don't have trust and respect on both sides, you can't have the hard conversations'. This perspective was also endorsed by farmers undertaking new developments – 'the right approach makes things go smoothly, and there are no unreasonable impediments from council'.

6.4.5 Global Relationships

Farmers were not convinced their future was as Asia's food bowl, though many saw opportunities in the growing Asian middle class, particularly in beef. A significant number were already involved in international trade and agreed 'the clean green image of Australia gives a premium no question, but premium on a fair price, not on an over-inflated domestic price'. And the need to pay attention to contractual arrangements was recognised, as 'Asians are very good business people'.

Foreign investment in Australian agriculture was a popular topic, but unlike some of the alarmist media headlines around foreign ownership, most farmers pointed out that there was no difference between today's Chinese money and the nineteenth- and twentieth-century British and American investors – 'they can't take the dirt home. Since Federation Vesteys have owned it, then Americans, Japanese, etc. People who complain are often demonstrably not good managers'.

While there have been spectacular examples of foreign investment gone wrong (see Chap. 3 "The Historic Challenges of Northern Australian Agriculture"), the American King Ranch development in Tully was described as 'the exception, and Robert Kleberg [the owner], at that time world's fourth richest man told my father that their success was because they sought the advice of local people already doing it', and King Ranch was seen as 'a benefit to all … the Americans got in and worked, leading by example'. Many believed that foreign investment was essential to northern Australia's agricultural growth, as 'Industry will be hamstrung if we rely solely on Australian investment … the dollars have to come from overseas'.

There was a warning though – 'The current push for an expansion of northern agriculture will be by corporate entities and this is the reason it will fail, as it will be by people without local knowledge making decisions from a central office'. Almost in acknowledgement of this, many talked of the opportunity and benefits from marrying Australian industry experience with foreign investment, and that possibly this was the key to a successful agricultural expansion in the north. Instances of this already happening were given – 'a Chinese entity has approached a large landholding in the area wanting to buy the land but leave it under current management for the next 20 years', though 'at the moment [there was] more speculation than real sales'. There were other warnings, including 'sometimes money can affect integrity' and 'around the world, big ag usually only returns enough to keep growers there, not to help them grow'. However, through the sale of the Tully sugar mill to a Chinese corporation 'Foreign investment has provided dignity for people to leave the land – we have to be grown up about this'.

Overall, farmers saw more opportunity than risk in foreign investment, though maintaining appropriate balances was important – 'we do need to retain control' was a common statement, as were suggestions like 'requiring majority Australian partnership'. However, there was recognition that 'Australia is not the only stable democracy in the world … Asia will look elsewhere if we don't do it', that 'not many Chinese are going to want to live out here … they prefer more settled areas' and that 'Northern Australia is more akin to a developing country than the rest of

Australia'. A key criterion for successful foreign investment and industry engagement was thought to be ensuring good human relations and maintaining reputation – 'people want a picture of you before they pick up the phone and ring' – strong bridging social capital.

Immigration was another topic raised with respect to the future of agriculture, and many pointed out that they were farming in Australia as a direct result of either their parents or grandparents emigrating from Asia, England, France, Ireland, Italy, Scotland or Spain:

> 'Australia is closing itself in … I wouldn't be here without immigration';
> 'Immigrants – yes! We've thrived on it, from the Italians to the Hmong. There's high numbers every year at nationalisation ceremonies … large numbers of Indians now, and they're good citizens'.

Part of this discussion related to contemporary limited labour availability, and part was concerned with maintaining the viability of regional communities. While today's backpacker travellers provide essential labour for farms and regional businesses, it is an itinerant population vulnerable to external influences such as adverse publicity or change in the value of the Australian dollar. Many farmers talked of the opportunities for regional population growth through the permanent settlement of refugees, particularly of 'people from rural areas who want to live here rather than wealthy people from cities'. Also, under current immigration policy, 'You can't sponsor someone into agriculture without a degree. There are thousands of categories, but the only two applicable here are manager or mechanic, not head stockman etc. It's not about resourcing, we need policy that encourages people to live out here'.

6.4.6 Relationship to Place

Despite mobility and globalisation processes, place continues to be an object of strong attachments for people, with almost 400 papers published within the 40 years prior to 2010 (Lewicka 2011). While the term *sense of place* has become something of a modern buzzword, used to describe everything from an appreciation of natural landscapes to the selling of home sites in urban sprawl (Cross 2001), the propensity of people to develop a sense of belonging, commitment, identity and community to, and through, where they live is universal. In fact, Faulkner (2014) (in Brown (2016, p. 123)) found that a sense of place was a critical precursor for community resilience and was rooted not just in physical space but among community and relations. The work of Cox and Perry (2011, p. 395) into communities affected by wildfire emphasises the 'critical importance of place not only as an orienting framework in recovery but also as the ground upon which social capital and community disaster resilience are built', and Hanna et al. (2009, p. 31) posit 'place can be seen as a materialisation of social capital'.

Australia's contemporary community is increasingly cognisant of the importance of place to its Traditional Owners, with the term *caring for country* concomitant with Aboriginal land management, and:

> 'Over the last few decades there has been a resurgence of effort by Aboriginal landowners to maintain their cultural responsibilities and knowledge, pursue socio-economic development opportunities, as well as to protect the bio-cultural values of their ancestral country'. (Moritz et al. 2013, p. 1)

Evidence of ever-more widespread acceptance of the concept was the Australian Government naming their 2008 $2.25 billion investment in sustainable land management *Caring for our Country*, with the intent to work with governments, regional and local communities, industries and land managers to achieve an environment that is healthier, better protected, well managed, resilient and can provide essential ecosystem services, particularly in a changing climate (Commonwealth of Australia 2008).

Throughout the interviews, farmers talked both overtly and sometimes through association, about the importance to them of where they lived and farmed – 'Being with country is very important ... not necessarily the Aboriginal context, but being close to the land – it's a very human thing', but was this in any way different to someone who has consciously chosen to live in inner-city Sydney because of the associated lifestyle and amenity attributes? Fortunately, it was never an intention to answer such a complex question, but the interviews did reveal that many farmers have consciously thought about where they live and why it is important to them and demonstrated that they value and respect both the place where they live and the community which they are part of despite any shortcomings or compromises that this choice necessitated:

> 'If I really wanted to make money, I'd move the whole operation to *[another property he owned south of the tick line and closer to transport and markets]*, but I've got an emotional attachment to this place and that's not going to change, so my market is live export and the meat trade'.

His wife obviously concurred with this strong sentiment, as she described telling the GABSI[21] team capping artesian bores near the station homestead 'if you don't get this right my garden will die, and if my garden dies then I will leave, and our marriage will end. No pressure, but …'

This commitment and obligation were beyond maintaining a productive landscape – 'the families [established in this district for more than 100 years] love the lifestyle, they love the industry, they love working with cattle, they love working with the landscape'. An older grazier from Queensland's western downs was considering retirement options and spent a few weeks in Townsville where he knew people and health facilities were readily available. At the end of that time, he told

---

[21] The Great Artesian Basin Sustainability Initiative is a joint programme between the Australian, New South Wales, Queensland, South Australian and Northern Territory governments to provide funding support to repair uncontrolled bores that threaten the long-term sustainability of the Great Artesian Basin.

me 'it's a nice place and I've enjoyed myself, but I couldn't wake up every morning and not see that big [western] horizon'.

This sentiment was not restricted to those born to their area: a Victorian-born resident of the Northern Territory described how from his first visit, he'd 'always been attracted to the cleanliness of the desert', while a Brisbane-born Far North Queenslander described how after his first visit at 18 years of age he 'fell in love with the place and the people' and decided he 'was going to buy some land and live here … it just took me 25 years'. While it is plausible that a strong attachment to a place could decrease resilience if a disaster forces relocation (Norris et al. 2008), it is equally plausible that those same attachments increase the likelihood that the individual and community has the will to rebuild after disaster, which was evident in the cyclone-affected areas of Far North Queensland.

Several pastoralists also discussed their responsibility to their livestock, and in some cases, this duty-of-care extended to pride, attachment and affection which won't surprise anyone who has visited the Brisbane *Ekka*[22] and seen the care lavished on stock. A grazier described his adult daughter's attachment to the property as 'because her animals are here and she wouldn't sacrifice them for a better social life'. A grazier talked proudly of her '52 years of breeding experience … you can't buy that again if it's sold' and then went on to describe a nearby farm tourism operation where 'He might have cash in the bank from his tourism, but his animals have to be looked after first, not last; and that's not happening. You have to be there when the animals need you, not when it suits you'.

The sentiment expressed in such instances was strong, much stronger than simply maintaining a financial investment. It was about responsibilities and obligations and morals, which is why one thought it essential for children to grow up with pets and for them to always 'feed the animals before they eat themselves, so they know their responsibilities [to animals]' – obvious commitment to the responsibilities of stewardship.

6.4.7 New Blood

> 'I did some orchard work in my youth and loved the lifestyle. They were big Italian families and included me in making their own wine, eating their own food … but land values make it hard for outsiders to enter *[agricultural]* industry ... Industry needs to talk about alternate finance options, but it's not happening'.

This interview participant was not a farmer – he works in an advocacy and policy role for a prominent agricultural industry body. He is articulate, and understanding of the diverse community views around farming and the environment and has been effective in bridging the gap between production and conservation perspectives in Queensland. He loves agriculture and says 'the best part of the job is kicking the dirt

[22] Queensland's annual agricultural show. Originally called the Brisbane Exhibition, it is commonly known as *the Ekka*.

with farmers'. Perhaps one day he will be kicking the dirt on his own farm? What was apparent was that while not born to a farm, he has an affinity with agriculture and is positively influencing its future.

All industry, in fact all life, benefits from innovation and interaction – from 'new blood'. There are many examples in agriculture, with improved breeds, clever mechanisation, and more sustainable farming practices, but Northern Australia starts from a low resident population and, in some regions, a declining farm demographic. Previously discussed was the circularity of needing more people to justify increased spending on infrastructure to, in turn, attract more people, but in some areas and industries, the more immediate challenge is to retain numbers.

Farming appeals to people as both a lifestyle and an industry – an extension officer 'grew up in Adelaide and went ringing because I loved riding motorbikes, found I liked the industry, so after 3 years went to Roseworthy Ag college as a mature age student', and numerous instances of people entering agriculture through their own initiative but assisted by others were recorded. A formal partnership whereupon a young couple bought equity in a new property with their previous employers has been described; and another told of a young fellow who worked for him for 3 years, loved the life, but didn't have 'the wherewithal to buy a place. But he married a nice girl, did his electrical contractors licence then went to Longreach where he did well'. He's now bought a block nearby and tells him 'his success was due to what I taught him; but he had a work ethic, ability and vision'. This relationship illustrates the two sides of the equation: earning the respect that convinces an established farmer to help. There were no examples provided of assistance being offered without some demonstration of commitment first, and a pastoralist observed that the opportunity for this to happen 'is declining – people don't go bush for a year like they did in [name supplied] day'.

Opportunities might arise in an expanding Northern Australian agriculture for locals skilled in particular industries to collaborate with foreign investors for mutual benefit. A possible mechanism for agricultural expansion supported by many in industry included:

> 'Corporate dollars + local expertise is the only way to make it [industry expansion] work. There're plenty of young guys around keen and enthusiastic who would love the opportunity. It's not a role for government facilitating this ... you need the willingness of the person to do it';
>
> 'Corporate grazing already does this – use experienced people, which provides a development process for people in the pastoral industry, but they progress people who do things the way they want them done';
>
> 'However, some observed that there is '*Risk in rolling out this model out as probably will get some free riders*' though this risk could be ameliorated through 'making managers more than managers through having a stake in the operation''.

Certainly, more is required than buying a farm and putting a manager on. A fruit grower observed that 'Many Asian agents [from the central produce markets] have bought tropical fruit farms, but they all seem to fail as they put managers on who don't understand the reality'. Another risk was that the 'chain of decision-making communication is vital, and this could be a risk if overseas owners need to approve important decisions like when to supplement, destock, etc.'.

Concerns were expressed about large corporate developments and what impact they would have on the regional community and industry, for example, 'Where are the opportunities for young farmers to get involved in something like IFED[23]? I asked [Member of Parliament] in a public meeting about the proposal and was publicly humiliated by him, but he didn't answer the question'. Notwithstanding these concerns, many participants believed future agricultural expansion will be through private endeavours rather than state-sponsored initiatives – 'North Australia hasn't been developed by government, it's been done by private companies in the main', and there will be opportunities in such expansion for both experienced and novice (but enthusiastic) participants.

New blood does not necessarily equate with small or niche farming: one of Queensland's large banana-growing families started in 1983 when 'Dad was an electrician at [a power station] which was automated in 1982, so he bought a farm'. However, it remains 'difficult to recruit professionals for regional areas' and 'hard to attract people [meaning professional and technical staff] with 3 year contracts … and there are no cadetships now' (see Chenoweth et al. (2013) for a discussion of the challenges and ethical dilemmas around professionals living and working in rural and remote Australia). A solution to this conundrum requires a whole of community response, as often 'The problem enticing professionals is usually wives and children – quality of life – and that is a role for Local Government. I tripled Parks and Gardens budget when elected and we won Tidy Towns. Perception is everything – have to look good' although 'Assistance to move to remote areas is not obvious in the current push. How to attract people without perverse outcomes? Chinchilla is offering residential blocks for $1, but will this work, or will it be a rural ghetto? Who actually pays for essential infrastructure?

6.4.8 Discussion and Conclusion

The importance to each person of where they live, how they lived their lives and how they interacted with others was evidenced throughout the interviews. The characteristics described align with one described by Brown (2016) as *rootedness* – which she uses to describe a person's identity and belonging through place, not just physical place but also among community and relations. Raymond et al. (2010, p. 433) also describe a 'valid and reliable measure of rural landholder attachments' through a five-dimensional model of place attachment, these being place identity, place dependence, nature bonding, family bonding and friend bonding.

Each of these dimensions were evidenced in my interviews, along with the additional (though linked) dimensions of business, and aspects of the business such as livestock. This supports the importance of *social connectedness* as a mechanism to realise change being a determining element of resilience. Lyon (2014) analytically disentangles the interwoven strands of a place and suggests that the physical, emo-

[23] See Chap. 3 'Contemporary Challenges to Northern Australian Agricultural Aspirations'.

tive and cultural elements of place shape social resilience in the face of crisis, and I am prompted to repeat the observation by Marshall et al. (2011, p. 154) that:

> 'Factors that make resource-users dependent on natural resources (such as attachment to occupation and place, education, employability, environmental attitudes, local knowledge, and the quality and extent of formal and informal networks) act to influence resource-users in their decisions to adopt strategies that could enhance their capacity to cope and adapt to climate variability'.

This concept is however 'relatively underrepresented in social ecological systems and human development literature' (Brown 2016, p. 123).

Fundamental to this theme of social connectedness is theory related to human and social capital, particularly the multifaceted conceptualisation of capitalised assets (Stokols et al. 2013). Contained within this theme are examples of research participants utilising bonding, bridging and linking social capital to assist them in managing the complexities of living their life and running a business, particularly when these two activities are intricately connected through remote geographic location and isolation. The value of moral capital was also strongly evidenced as a critical determinant of that lubricant of social life – trust (Putnam 2000), whose importance was also enhanced through the effects of distance and low human population. The over-riding sense from this theme was the importance of self-efficacy to the individuals, articulated through self-belief in their own capacity, that is, their personal agency.

References

Adger, W. N. (2006). Vulnerability. *Global Environmental Change, 16*, 268–281.

Agriculture Victoria. (2016). *Whole farm planning*. Retrieved from http://agriculture.vic.gov.au/agriculture/farm-management/business-management/whole-farm-planning

Brennan, M. A. (2008). Conceptualizing resiliency: An interactional perspective for community and youth development. *Child Care in Practice, 14*(1), 55–64. https://doi.org/10.1080/13575270701733732.

Brown, K. (2016). *Resilience, development and global change*. London: Routledge.

Brown, K., & Westaway, E. (2011). Agency, capacity, and resilience to environmental change: Lessons from human development, well-being, and disasters. *Annual Review of Environment and Resources, 36*, 321–342. https://doi.org/10.1146/annurev-environ-052610-092905.

Chenoweth, L., McAuliffe, D., Tracey, P. J., O'Connor, B., Klieve, H., Stehlik, D. (2013). *Ethical dilemmas of everyday rural life: How do professionals balance living and working in rural and remote Australia*. Paper presented at the Inter-Disciplinary. Net 3rd Global Conference, Prague. https://experts.griffith.edu.au/publication/n940e098028dee5b8513e693076c6b847

Commonwealth of Australia. (2008). *Caring for our country: Outcomes 2008–2013*. Retrieved from www.uq.edu.au/agriculture/docs/CaringForOurCountry-Outcomes.pdf

Cox, R. S., & Perry, K. E. (2011). Like a fish out of water: Reconsidering disaster recovery and the role of place and social capital in community disaster resilience. *American Journal of Community Psychology, 48*(3–4), 395–411. https://doi.org/10.1007/s10464-011-9427-0.

Cross, J. E. (2001). *What is sense of place?* Paper presented at the The 12th Headwaters Conference, Western State Colorado University. Retrieved from http://western.edu/sites/default/files/documents/cross_headwatersXII.pdf

Darnhofer, I. (2014). Resilience and why it matters for farm management. *European Review of Agricultural Economics, 41*(3), 461–484. https://doi.org/10.1093/erae/jbu012.

Darnhofer, I., Fairweather, J., & Moller, H. (2010). Assessing a farm's sustainability: Insights from resilience thinking. *International Journal of Agricultural Sustainability, 8*(3), 186–198. https://doi.org/10.3763/ijas.2010.0480.

Dowd, A.-M., Marshall, N., Fleming, A., Jakku, E., Gaillard, E., & Howden, M. (2014). The role of networks in transforming Australian agriculture. *Nature Climate Change, 4*(7), 558–563. https://doi.org/10.1038/nclimate2275.

Elder, G. H. (1994). Time, human agency, and social change: Perspectives on the life course. *Social Psychology Quarterly, 57*(1), 4–15.

Eriksen, S., Aldunce, P., Bahinipati, C. S., Martins, R. D., Molefe, J. I., Nhemachena, C., et al. (2011). When not every response to climate change is a good one: Identifying principles for sustainable adaptation. *Climate and Development, 3*(1), 7–20. https://doi.org/10.3763/cdev.2010.0060.

Faulkner, L. (2014). *Assessing community resilience in North Cornwall: Local perceptions into responding to changing risk landscapes.* (M.Res dissertation), University of Exeter.

Folke, C. (2006). Resilience: The emergence of a perspective for social–ecological systems analyses. *Global Environmental Change, 16*, 253–267.

Folke, C., Colding, J., & Berkes, F. (2003). Synthesis. Building resilience and adaptive capacity in social-ecological systems. In F. Berkes, J. Colding, & C. Folke (Eds.), *Navigating social-ecological systems. Building resilience for complexity and change* (pp. 352–387). Cambridge, UK: Cambridge University Press.

Gallopín, G. C. (2006). Linkages between vulnerability, resilience, and adaptive capacity. *Global Environmental Change, 16*, 293–303.

Goldsmith, E. (1978). *The stable society*. UK: Edward Goldsmith.

Golomb, J., & Wistrich, R. S. (2002). *Nietzsche, godfather of fascism?: On the uses and abuses of a philosophy*. Princeton: Princeton University Press.

Great Barrier Reef Water Science Taskforce. (2016). *Great barrier reef water science taskforce – final report*. Queensland Government. Retrieved from http://www.gbr.qld.gov.au/documents/gbrwst-finalreport-2016.pdf

Hanna, K. S., Dale, A., & Ling, C. (2009). Social capital and quality of place: Reflections on growth and change in a small town. *Local Environment, 14*(1), 31–44. https://doi.org/10.1080/13549830802522434.

Holling, C. S., Berkes, F., & Folke, C. (1998). Science, sustainability and resource management. In F. Berkes, C. Folke, & C. Béné (Eds.), *Linking social and ecological systems. Management practices and social mechanisms for building resilience* (pp. 342–362). Cambridge, UK: Cambridge University Press.

Kuhlicke, C. (2010). The dynamics of vulnerability: Some preliminary thoughts about the occurrence of 'radical surprises' and a case study on the 2002 flood (Germany). *Natural Hazards, 55*(3), 671–688. https://doi.org/10.1007/s11069-010-9645-z.

Lewicka, M. (2011). Place attachment: How far have we come in the last 40 years? *Journal of Environmental Psychology, 31*(3), 207–230. https://doi.org/10.1016/j.jenvp.2010.10.001.

Lorenz, D. F. (2013). The diversity of resilience: Contributions from a social science perspective. *Natural Hazards, 67*(1), 7–24. https://doi.org/10.1007/s11069-010-9654-y.

Luthar, S. S. (2006). Resilience in development: A synthesis of research across five decades. In D. Cicchetti & D. J. Cohen (Eds.), *Developmental psychopathology* (Risk, disorder, and adaptation) (Vol. 3, 2nd ed., pp. 795–739). New York: Wiley.

Lyon, C. (2014). Place systems and social resilience: A framework for understanding place in social adaptation, resilience, and transformation. *Society & Natural Resources, 27*(10), 1009–1023. https://doi.org/10.1080/08941920.2014.918228.

Magis, K. (2010). Community resilience: An Indicator of social sustainability. *Society & Natural Resources, 23*(5), 401–416. https://doi.org/10.1080/08941920903305674.

Marshall, N. A., Gordon, I. J., & Ash, A. J. (2011). The reluctance of resource-users to adopt seasonal climate forecasts to enhance resilience to climate variability on the rangelands. *Climatic Change, 107*(3–4), 511–529. https://doi.org/10.1007/s10584-010-9962-y.

Masten, A. S. (2001). Ordinary magic: Resilience processes in development. *American Psychologist, 56*(3), 227–238. https://doi.org/10.1037/0003-066X.56.3.227.

Montagu, A. (1942). *Man's most dangerous myth: The fallacy of race*. New York: Columbia Univ. Press.

Moritz, C., Ens, E., Potter, S., & Catullo, R. (2013). The Australian monsoonal tropics: An opportunity to protect unique biodiversity and secure benefits for Aboriginal communities. *Pacific Conservation Biology, 19*(3/4), 343–355.

NATO Science for Peace and Security Series. (2007). *Environmental change and human security: Recognizing and acting on hazard impacts.* Paper presented at the the NATO Advanced Research Workshop on Environmental Change and Human Security: Recognizing and Acting on Hazard Impacts, Newport, Rhode Island.

Nelson, D. R., Adger, W. N., & Brown, K. (2007). Adaptation to environmental change: Contributions of a resilience framework. *Annual Review of Environment and Resources, 32*(1), 395–419. https://doi.org/10.1146/annurev.energy.32.051807.090348.

Norris, F. H., Stevens, S. P., Pfefferbaum, B., Wyche, K. F., & Pfefferbaum, R. L. (2008). Community resilience as a metaphor, theory, set of capacities, and strategy for disaster readiness. *American Journal of Community Psychology, 41*(1–2), 127–150. https://doi.org/10.1007/s10464-007-9156-6.

Putnam, R. D. (2000). *Bowling alone: The collapse and revival ofAmerican community*. New York: Simon & Schuster.

Raymond, C. M., Brown, G., & Weber, D. (2010). The measurement of place attachment: Personal, community, and environmental connections. *Journal of Environmental Psychology, 30*(4), 422–434.

Richardson, G. E. (2002). The metatheory of resilience and resiliency. *Journal of Clinical Psychology, 58*(3), 307–321. https://doi.org/10.1002/jclp.10020.

Smit, B., & Wandel, J. (2006). Adaptation, adaptive capacity and vulnerability. *Global Environmental Change, 16*, 282–292.

Stokols, D., Lejano, R. P., & Hipp, J. (2013). Enhancing the resilience of human–environment systems: A social ecological perspective. *Ecology and Society, 18*(1), 7. https://doi.org/10.5751/ES-05301-180107.

Wisner, B., Blaikie, P., Cannon, T., & Davis, I. (2003). *At risk second edition: Natural hazards, people's vulnerability and disasters*. London: Routledge.

Chapter 7
The Influence and Importance of the Built Environment

When it's hot, we find a shady tree to sit under; when it's cold, we find a windbreak and build a fire; and if there's danger, we find a safe protected place. This is what people have always done; we're just a little more sophisticated in the twenty-first century. It's still about function and amenity of the shelter: *function* – the requirements of utility and economy, efficiency in construction of a serviceable building for the needs of production and profit. If function brings short-term advantage, *amenity* brings long-term satisfaction – how a building or a place works for its user, its current and future purpose. A building that is open-ended and changeable can more easily be repurposed or repaired after a disaster. The eye and movement through these spaces play a crucial role in our visual and tactile experience of place.

Architecture is the consequence of human interactions and lies on the realm of memory and the experienced. Architecture has the power to comfort and to inspire, and through the intimacy of places and moments lived, there is a blurred line between the built and natural environment – dream and reality – time and space. Every creative endeavour is woven from and relative to the processes that influence the way we experience the world, and mostly it's the little unnoticeable things that create sweeping change.

7.1 Introduction: A Place to Call Home

Home for me is a place that I can feel comfortable, healthy and happy with the people I love, doing what we like, learning, giving, caring and sharing – a place of wellbeing, security, solace and, at times, metamorphosis. Home is private and semiprivate space – a place to sleep, eat and live. A garden. The longer people can stay in their homes, the longer they stay in their community, the greater the bonds of trust and social

---

Tania Dennis is the primary author of this chapter.
Tania Dennis and Keith Noble have contributed more to this chapter.

K. Noble et al., *Agriculture and Resilience in Australia's North*,
https://doi.org/10.1007/978-981-13-8355-7_7

connection. A home is a single collective unit inside the social machine of a community. There's the here and the now, then there's memory and history, and it all contributes to how I understand my local and global place in the world – all are important in my daily face-to-face interactions with life, my environment and my neighbourhoods. It's my personal space and part of who I am – not as an architect, as a person.

I was born in Darwin and raised in Katherine, and flying with my Dad to install and service generators across remote Top End communities and stations nourished my dreams and provided intimacy with life in the North Australian tropics. Through these influences, I learnt about people, place and buildings in remote and beautiful landscapes. Often hard landscapes, hard lives and stoic people; on the surface at least, until you got to know them. This time taught me enduring life lessons and to always respect the interdependence of place, experience and its people – and that the whole point of the tropics is living outside.

People in rural Australia have a sense of belonging and interdependence, and by shaping their environment, they contribute to and connect to their community. Being smaller than cities, most rural communities are cohesive and actively engaged, strengthening residents' sense of belonging and participation. The fabric and resilience experiences of my life are bound in the connection between architecture and life and a fundamental premise of my practice – architecture is philosophy. The built environment influences the people who use and experience it just as much as they influence the environment in which it is built.

I recall, as a 7-year-old, arriving at my grandparent's house that they had built into the cliff on the Darwin Esplanade, surrounded by shady trees and with big prop shutters to maximise air flow through the house. I remember walking down to the laundry entrance and seeing my Nana hand-washing clothes on a glass washboard. That night Cyclone Tracy changed a lot of people's values, including mine.

7.2 What Is Architecture and Why Is It Important

City skylines, imposing public buildings, funky 'Grand Design' houses … sure, architects do that, but they can also do a whole lot more. Architecture is a process of empowerment. Architects can help us, as individuals or communities, distill the vast collective human knowledge and experience of a place to live work and play and focus it on a location or aspiration to realise a vision. Architects aren't magicians – they're conductors, aligning and harmonizing the skills, knowledge and expertise required to turn dreams into reality irrespective of scale, and it is as much a process of empowerment, building capacity and nurturing hope as it is about creating spaces that nurture people. Good places aren't designed *for* the community; they're designed *with* the community. A good architect brings people together around an idea, but not from the top – they inspire collaboration; and codesign more often enables adaptable places that can respond to change of use.

What a place looks like matters, but not in a *keeping up with the Jones* sense – the character, design and quality of a neighbourhood influence people's sense of belonging in their community and contribute to social interaction and intercultural understanding. And design is not necessarily the most important element in the development of a good built environment: the folding together of practicalities needed or preferred use of public space can result in a unique design – witness the distinctive individuality of many of the world's great multilayered cities.

Architecture can contribute positively to our overall quality of life, health and wellbeing. Through good design, buildings can inspire, grow and adapt to suit changing needs, and buildings and places should be beautiful, functional and sustainable. Architecture has a significant influence on wellbeing through shaping our individual experiences in a place, and collaboration lies at the heart of all our work. Good architecture is too complex for one person: the process of briefing, designing, constructing, operating, using and maintaining buildings has to begin with identifying and then understanding the client's business case, strategic brief and other core project requirements; and never assume that the first thing people tell you is really what they want. So, we ask (Fig. 7.1):

This conversation is not particularly easy when everyone speaks the same language and has at least some shared societal/cultural/life experience, so you'd think it would be especially difficult when there was no shared language, cultural orientation or familiarity with basic building concepts and functions and for a project in an extremely isolated and demanding physical environment, with very limited access to experienced building service providers. Not so.

Fig. 7.1 What works best for you? (Courtesy of Insideout Architects and Mark Thomson)

7.3 Flying Woman Creek: Local Knowledge

Driving out to Kalkatarra with a Toyota full of Ngaanyatjarra women to check some rock hole ideas for *Tjulyuru*, we crossed sand dunes and occasionally stopped to burn country. We crossed the final sand dune, then a small stony rise with granite boulders strewn all over it, and the country swooped down, and you could see purple ranges in the distance. Driving down the other side of this rise, someone in the back of the Toyota called out 'watch the creek' and I thought okay, so where's the creek? ... BOOM! We hit the creek and flew off our seats. Ngaanyatjarra women call this creek 'Flying Woman Creek' because most people drive through it at a great rate of knots like I did. Local knowledge, you can't beat it.

I'd gone to Warburton community to build a cultural and civic centre for the Shire of Ngaanyatjarraku. At 400 people, Warburton was the largest of the 11 Ngaanyatjarra communities, where 2000 Yarnangu[1] live – the traditional and uncontested owners of 3% of mainland Australia. The existing design was by a big-city architect who had never visited the lands and, I discovered, would require a budget more than twice that available. In addition, the community's power station would require a significant upgrade if we were to build it and then turn the air-conditioning on. So when Ngaanyatjarra people asked if I could build something with the money available, I said 'let's try, together' – so for the next 3 years, I was Australia's most isolated architect.

Warburton was established by the United Aborigines Mission in 1933. People still lived wholly traditional lives till as late as the 1980s; most Elders were born in the bush; and English was very much a second language. Their experience with built forms was predominately limited to an imposed *cost-effective* housing system. Their answer to the question 'what sort of house would you like' was obviously 'the same as everyone else', as neither the language nor built form understanding existed to specify anything else. In fact, the whole concept of living in one place was still a novel and largely untested concept, with only two or three generations having done so compared with the hundred thousand prior generation's experience of living with and of the land. But Ngaanyatjarra are an adaptable and adventurous mob, and they understood that a civic centre was just what they needed to better interact with the rest of the world.

The initial consultation process was slow as we learnt to work together in this cross-cultural environment. We designed not from pictures but from words, notes and models: who will be there; when will you be there; what will you do there; where will you sit; where will you eat; where does the wind come from in winter; and where is the shade in summer? Where will the kids play and who will keep an eye on them? We made boxes for places, with price tags, and Ngaanyatjarra people went shopping, arranging various layouts of their building, gardens and courtyards until they got something that worked for everyone.

While we were codesigning Ngaanyatjarra Tjulyuru, we also worked out how Ngaanyatjarra people wanted to run the place and manage tourism in the lands. We

[1] The Aboriginal people who speak the Ngaanyatjarra dialect and live on their traditional homelands in the western central desert of Australia.

drove to rock holes, *reera*[2] and gorges where we dug for *tirnka*[3] and *yirlirltu*[4] and talked about what was special about those places, when was a good time to visit them, how the colours changed with time of day and seasons and discrete connection to country. And I discovered Flying Woman Creek. Because I wore trousers and kept my hair short, I was considered a somewhat androgynous being, so I could talk to the women and men. I was neutral. Not that we needed any of that secret stuff – just an understanding of how people liked to live their life on a day-to-day basis. And what was important to them.

This Ngaanyatjarra space, *Tjulyuru*, is one in which the sacred and temporal are fused, and this amalgamation is the underlying concept for both the built form and its landscape. The architectural form of the building and its extension into surrounding places through the landscape has many references to Ngaanyatjarra culture and society, and the site itself is a significant cultural location, but I didn't need to know the details. The secret stuff, well that was mostly kept quiet, though there are discrete references through landscapes and vistas for the people who knew where to look.

Construction in one of the most remote locations in Australia was not easy. There are a lot of people to manage in a project, and there are a lot of people behind a good building. Working alongside people in remote areas to deliver their dream does take a certain kind of tenaciousness and passion. It doesn't matter who or what you are, you have to love what you do. A mix of Ngaanyatjarra and outside contractors were employed and considerable logistic hurdles and cultural issues addressed. The project was completed within budget and 9 weeks early. The result is a well-considered living architecture, bridging non-Aboriginal and Aboriginal culture. The Tjulyuru Cultural and Civic Centre is the home and exhibition space for the nationally recognised Warburton Art Collection, some 300 paintings and architectural art glass. This is the most substantial collection of Aboriginal art in the country still under the direct ownership and control of Aboriginal people. The community's desire to retain this fundamental cultural material was central to the project.

Tjulyuru has a big internal courtyard with rock holes, ground patterns, trees, places for fire, sitting and talk or cooking quandong pies (or Fray Bentos® Steak & Kidney pies), wattle seed damper, working on paintings or making artefacts. It is woven into Ngaanyatjarra people's front yard along the Great Central Road, which links Perth to Cairns via Uluru. The Centre is a container of aspects of *Ngaanyatjarra* (this/separate/people) and will immerse people in a significant mutual space where they can have an immediate and compelling experience of Ngaanyatjarra culture and society. In 2000, it was awarded the Northern Territory's most prestigious Architectural Award, the Tracy Memorial Award, along with the Commercial Award and the Environment Award for energy-efficient design. But more important than these awards was a Ngaanyatjarra woman saying to me, 'Thank you for listening to us and doing things the right way' (Figs. 7.2 and 7.3).

---

[2] Undulating stony plains, largely free of spinifex or trees.

[3] Goanna.

[4] Honey ant.

Fig. 7.2 Tjulyuru courtyard. (Photo by Tania Dennis, courtesy Insideout Architects)

Vignette 7.1: Tjulyuru

FAR from Australia's great cities, like a gleaming secret unknown and unseen by outsiders, lies the nation's deepest, most coherent and focused collection of desert Aboriginal art. On public view for the first time over the next 2 months at the Tjulyuru Regional Arts Gallery in the remote heart of the Victoria Desert, these key works from the Warburton Collection complete our map of Western Desert painting.

New ways of portraying the desert are on offer here and they require new forms of appreciation and response … What was clear at once, though – from the colours, from the subjects – was that this art had not been conceived for outside consumption. The colours hung together in striking, unfamiliar fashion; there were paintings that, to Western eyes, seemed ridiculously full of water snakes.

The Warburton Collection is the nation's largest private assemblage of desert art collected and retained in indigenous hands. It has a distinct role, as an underpinning of traditional Ngaanyatjarra society. But what part could it play in the modern life of the desert? Could it be a bridge to the wider world?

Fig. 7.3 In Tjulyuru Gallery with Lalla West and Tania Dennis. (Photo by Tania Dennis, courtesy Insideout Architects)

The shimmering circles and the wavy, mirage-like lines of the canvases hang mute on the Tjulyuru Gallery's walls, like so many echoes of the desert landscape stretching away through the windows outside. They mirror, of course, the country: it gives them their power, and they take their force from its austere, receding splendour. A visit to the gallery, in fact, is enough to set a disturbing thought in mind: out here amid the rangelands of the Gibson Desert, desert art seems to gain a dimension, the depth of place. City galleries and museums of Aboriginal paintings, however subtly conceived, inevitably have something in common with the great museums of Western art, so full of religious paintings stripped from their altars, offered up as mere works of beauty, shorn of spiritual charge.

Hidden Testament
By Nicolas Rothwell
The Weekend Australian, 08 May 2004

7.4 The Classical Elements

Architecture is design for wellbeing and is about understanding how building elements can improve the health and wellbeing of its occupants, from air, water, light, materials to space planning, fit-out and gardens. These classical elements are reflected in many cultures as they provide a human context to buildings and places (see Alexander et al. (1977):

Earth – life-giving, creator, land, connecting to and looking after country – cartographers landscapes, cities and towns – to define, explain and navigate our way through and around the world.

Life – balancing ecological harmony, sustainability and diversity. Contributing to local biodiversity through human-inclusive biophilic design principles. Recognising and nurturing the human/nature connection and interactions, both inside and out. Transforming projects by incorporating natural shapes, forms, light, space and air and indoor gardens that improve air quality. Connection to place, climate and culture through place-based relationships that develop outside gardens, courtyards, seats and verandahs. Ecologically restorative use of local building materials that contribute to biodiversity and are compatible with the landscape.

Water – optimum water quality and efficient use of water inside and in gardens – water for drinking, food preparation, bathing and washing. Permeable paving in courtyards or driveways that house underground water storage.

Fire/light – maximising internal and natural daylight to minimise the need for artificial lighting benefits natural body rhythms. Orientation and opening arrangements to optimise light, views and connection to nature.

Air/wind – cool fresh air through good ventilation principles. Openings that capture prevailing breeze and use of non-toxic locally sourced materials wherever possible.

Aether/sky – inspiring places that enrich our lives and experiences – a canvas for creativity and happiness. Spaces that create atmosphere and theatre in the building

Then, by looking at a building's context you can innovatively design for adaptability and longevity – places that are *loose* are more likely to adapt or expand to suit evolving needs within lifetimes and across lifetimes. Designing and building for wellbeing means including places to discover, learn, research, be active in, get connected, give and explore and don't necessarily cost a lot of money. It's remembering to connect and to think beyond today (Fig. 7.4).

People: connect with people – provide places to gather, share a meal or a conversation – bright green leafy verandahs and shady sheltered courtyards. Get together with friends and family and share a meal under the sky. A relaxed tropical lifestyle, with balance in work, life and play.

Fig. 7.4 Different. (Courtesy of Insideout Architects and Mark Thomson)

Self: breathe – be active – keep yourself nice. Keep your place nice. Provide accessible buildings with universal equitable access.

Place: innovate, and create connection with the bigger community; reflect our role and connection. Livable connections with interwoven private and public space. Shared places, internal and external, for living. Positive environmental and health outcomes through the architecture. Convenient. Spacious. Pause. Reflect. Serene. Lively.

Mind: keep learning; listen. Strengthen community. Seek out and respect local knowledge. Multicultural heritage. Memory. Diversity. Opportunity. Adapt. Grow.

Spirit: meet, give, let life flow. Embrace local culture and art. Seek out places to share and listen, spaces where people can feel at ease and retreat from their daily stresses to regenerate and nurture their spirit and connections with nature.

Landscape: we live in landscapes – with aesthetics and connectivity to other immediate and wider landscapes – care and balance of urban with green – about your home and yard, and extend this into your community. Nature supports human wellbeing and people support nature in a mutually beneficial way. Use locally sourced natural materials. Create imaginative inspiring outdoor learning places. Using local skills, knowledge and talent brings together people who appreciate and value a healthy life and provides awareness of the importance of conservation for our own wellbeing and that of the planet.

Vignette 7.2: Lightning Through the Floorboards
Tropical Cyclones don't travel in straight lines, they change course; just like life.

December 24, 1974: On our drive from Katherine to Darwin there was a loud BEEP, BEEP, BEEP out of the Land Rover radio … 'Be quiet you kids so we can listen to this'…

'PRIORITY! CYCLONE WARNING!'

… we were listening to something nobody wants to hear… 'SEVERE TROPICAL CYCLONE TRACY IS CENTRED 80KM WEST-NORTHWEST OF DARWIN AND MOVING SOUTH-EAST AT 7 KILOMETRES PER HOUR. THE CENTRE IS EXPECTED TO BE NEAR'.

… Dad said 'It looks like we're in for a blow'. We continued to drive – to see our relatives and my grandmother in her 1920s prop-shutter house on top of the Darwin Esplanade cliff face, overlooking the Arafura Sea. We were kids, and it was Christmas.

MIDNIGHT CST. 24 DECEMBER 1974: By now we could barely hear the warnings over the wind and waves lashing the cliff below my Grandmother's house. We were downstairs now – my Pappoús had dug the house into the sloping ground at the top of the cliff, so it felt like we were sheltering in a bunker, my Nana saying 'This house has been through the bombing [of Darwin during World War 2] and survived many cyclones – take the children upstairs so they can sleep my dear' and Dad saying 'they can't, the roof has gone' – Nana not believing until she saw the lightning through the floorboards.

7.5 Other Elements

The catastrophic experience of Tropical Cyclone Tracy in 1974 introduced me to the temporal nature of architecture and has been with me ever since – the need to have structures that are comfortable and durable – and of the people and their place. Initial response to this destruction was 'an over-reaction to the cyclone and climatically inappropriate. Architects just didn't seem to be looking at their own local building traditions. They were ignoring an entire heritage of Top End architecture' (Goad 2005) (Fig. 7.5).

Cyclone Tracy blew Darwin away and changed a lot of people's values, including mine. People mattered, and material things didn't. Tracy was a real leveller because everyone who survived was in the same boat. Some people made a fresh start, while others left and never returned. Darwin community worked as a collective to transcend destruction and rebuild a different Darwin. Darwin society changed

Fig. 7.5 Early Darwin living – light-weight and tropical. (Photo by Jessie Litchfield, courtesy the Tania Dennis Collection)

and so did people's dream homes. The chaotic disorder created by a natural disaster introduced a new imposed order on Darwin – the desire for resilience. The concrete box design mentality took over – a concrete *esky*[5], strong and durable, but a climatically impractical solution designed and built solely on the functional need for survival. Practically and culturally, the concrete *esky* didn't work in the tropics; and mango madness was ripe. Though still young, I remember when Troppo Architects came to town with their 'Green Can' house that, just like my grandmother's house, referenced early Darwin homes from the 1880s to 1920s and used tropical construction elements such as timber slats, woven or bamboo walls and shutters and vegetation.

So, when it came to designing a tropical open house in a cyclone region, this came naturally to me. Brooks Beach House is a strand of pavilions perched on the rim of a very steep rainforest slope some 150 m above the ocean. The house, with the fragile rainforest habitat of the critically endangered cassowary, aims to minimise environmental impacts. This tropical house combines working and living space for two professionals, with specific and contrasting physical requirements. Together with the engineer, steel shop detailer, steel fabricator and builder, we detailed a house that was climatically responsive, minimised construction time and costs, maintenance and cyclone exposure. Completed in 2003, Brooks Beach House was the only house in the area that largely withstood severe Category 5 tropical

[5] A portable cooler you fill with ice to keep things cold.

Fig. 7.6 Brooks Beach House, Brooks Beach Queensland, Australia – in its setting
'As the architect works organically she was able to feed into and be challenged by the client's wildest dreams. The result is a series of structures that are adventurous, dramatic, bright and playful. The architect by her own admission is a modernist, yet responded with vigour to the clients demand for more and more chaos. This house clearly reflects the diverse personalities of the two clients. It is about performance and narrative culminating in a dialogue that is articulated by the play of the pavilions. It is nothing short of delightful!' Jury Statement, Australian Institute of Architects. (Photo by Mike Gillam, courtesy Insideout Architects)

cyclones Larry (in 2006) and Yasi (in 2011) that both crossed the east coast over Brooks Beach House, while the owners sheltered inside. Although light-weight and tropical, the house was durable through honouring the ABC construction principles – anchorage, bracing and continuity (Fig. 7.6).

Can design tinker[6] with the social architecture of our minds? A diverse range of flexible housing forms ensure places can accommodate people from different backgrounds, and they can expand or modify the composition of their home as their needs change. Common spaces foster healthy interactions among people with different interests. Private to public – verandahs and gardens – social spaces create a clear though gentle transition between private and public realms. Strong social connections can emerge when people have opportunities for unscheduled interactions with neighbours, knowing they have somewhere to retreat if the mood takes them, maintaining their agency by providing personal territory, safety and satisfaction.

[6] 'I tinker, therefore I am'. Mark Thomson, Institute of Backyard Studies.

Disasters like tropical cyclones can shake our communities into better places 'destroying a community's past mistakes to make a community more livable and healthy … an opportunity to build it better' (Donovan 2013). Disaster engages an ethnically and socio-economically diverse community and can bring them together. Shocks or disasters can shake us out of our habitual reserve, and people from other communities come to help and support those in need, and everyone learns from the experience. We share responsibilities, resources and activities that give flexibility and support to one another during post-cyclone clean-ups. We realise we are not alone and that there is genuine good in people and communities.

Outside the context of disaster, can we change our communities using the knowledge of these experiences? Strangers working together – strong social connections emerge when people work or play on tasks or causes bigger than themselves, e.g., social and sporting clubs and community gardens. Such strategies enable communities to meet a range of interests and needs, and the interaction of people of different ages, ethnicities, income levels and household sizes develops social opportunities and multigenerational support.

Local knowledge – local community-based knowledge – is vital post-disaster, when locals guide disaster workers as they identify vulnerable homes and public spaces suitable as temporary service centres, negotiate or preserve cultural complexities and appropriately manage natural environment systems[7]. This is where, as a community, we get back to the specifics of place and approach design using local knowledge and an insider understanding of how things work based on local experience, rather than on untested theory or the experience from elsewhere. Increasingly, designers are aware of the importance of local knowledge as a dimension of healthy community life and maintaining it. Maintenance of local knowledge comes from informal social interaction through public spaces where people can learn about each other and share their knowledge. Local knowledge is fostered by publicness – by people connecting with each other in face-to-face interactions because they're visible.

As architects, we learn from natural disasters. Darwin's post-Tracy aesthetic opened designers to a tropical model of design with permeable walls, connection to the outdoors and passive coolness; or did the cyclone shake off an imposed non-regional mindset to rediscover traditional resilient design and construction techniques used in disaster-prone regions around the world – similar methods because they work? If we accept that climate change and natural disasters are going to happen, and if we live with and work with them, they are less likely to scare us into inappropriate responses when they do occur.

[7] The emergency clearing of creeks and other important habitat after Cyclone Larry was of concern to many residents. Drawing on this experience, environmental clearing post-Cyclone Yasi was managed in association with the Regional NRM body *Terrain*, with excellent outcomes.

Vignette 7.3: Shed (Shared) Living

When they arrived, they found the tent had been pitched under the shade of a huge kurrajong tree. It was still growing green and strong in front of the new homestead, standing out proudly among the poincianas and oleanders and frangipani and all the other Johnny-come-lately foreign trees about it. She could remember lying on her back on the short thick grass in front of the tent and looking up into the dense canopy of leaves above her and seeing a tiny red-backed wren peeping at her round a twig. His little head was cocked sideways and his perky spike of a tail jerked energetically. She pretended he was beckoning her into his house and she was suddenly happy and at home.

It had not been much of a home, by city standards; only the most primitive of accommodation. A couple of tents and a bough shed under the trees, then after a year, a single roomed hut with paperbark roof and adobe walls. As time went on, tin had replaced the bark on the roof, and two verandahs had been added and a little kitchen annex with its huge stone fireplace. The present homestead, was comparatively recent. They used the old house as a barn now, and sometimes Sonny would stand there among the feed bins and the great racks of stored grain, and remember where the kitchen table had stood up against the east wall, and how the little fat geckoes had come out of the cracks in the adobe and scuttled over the table at night after the insects that fell around the light. (They fell onto the table off the ceiling in the new house – once one had landed in the soup with a tremendous splash!)

Mum and Dad had slept over against the west wall, and every time it rained a single drip would land right on Dad's head. He had spent hours looking for that leak, but he could never find it. Still, they had to have the nets up in the wet, and Mum just put a dish in the right place on top of the net, so that took care of the drip. She could still hear the small musical 'plink…plonk' as the water dropped slowly into the tin dish.

Wallaby Man by Helen Litchfield (2012) – an extract from her unpublished manuscript describing her 1920s home on the Roper River.

7.6 A Sense of Place

The tea things were put out on the small cane table on the north verandah, and everyone just helped themselves and sat wherever they pleased. There were a couple of squatters chairs and a couple of home-made easy chairs of unique design, their shape dictated by the curve of the tree from which they had been built. They were wide, deep and comfortable, and so large that Sonny practically disappeared from sight in them – exactly what she wanted to do today. (Litchfield 2012)

People feel a stronger sense of comfort and attachment to places that reflect their culture, values and sense of self. Local aesthetics reflecting local identities increase satisfaction and support social wellbeing by allowing self-expression and individuation. Collective values can be nurtured by providing spaces for people to share ideas, work on new projects and support each other. Spaces for cocreation and self-expression help define a collective identity in a place, and people are more likely to talk to one another while they're cocreating. Working together is a powerful generator of social trust and happiness – learning by doing, design in the real world, looking to the future, building, testing, refining – reframing architectural research and construction design language redefines community values and commitment to make things work. Good design is for the whole of society, not a single client, and good codesign feels like a gift. When a community participates in the planning and design process, social relationships are strengthened, and trust grows between all players.

People are happier in their homes when they have a visual window connecting them to nature. Gardens promote physical and mental wellbeing, and green space is one of the strongest correlates of health and happiness. When people enjoy direct contact with nature, they are more likely to appreciate their environment and engage in sustainable living, and controllable natural light increases satisfaction and comfort; and when people are comfortable, they are more likely to engage with others.

Spaces and places have memory – physical artefacts help us remember things. Every time I go back to Darwin, I remember Tracy and build on my knowledge of what happened in the past. The destruction of Darwin removed the collective memory of Darwin's people who chose to rebuild it again – architectural behavioural shapers, environmental determinism – driving positive change through the power of design, the human response to environment.

There is a difference between a building triggering a physiological response and architecture changing behaviour. We are products of our choices more than of our environment. People make places, and as we become an increasingly global connected society, it's important to resist becoming international and anonymous. Homes and cities are behavioural devices – their shapes and systems alter how we feel, how we see each other and how we act. This would be a terrible thought if it were not for a second truth, which is that our homes are malleable – we can change it whenever we wish. We can redesign our homes, our minds and our own behaviours to build a life that is easier on the planet, more convivial, more fair, more fun and more happy. When people travel, they are open to new ideas, new tastes and new experience. People leave home to see things and are intrigued by and drawn to places different from their own – to experience someone else's sense of place. And we like to bring home a souvenir, a little something to remember it by – a tea towel or a change to our home inspired by something seen or experienced, which brings us to the Ghunmarn Cultural Precinct in Wugularr community, 100 km east of Katherine, where cultural tourism is helping drive regeneration and growth.

7.7 The Space Between Is a Collaboration Machine

The space between buildings invites occupation and resonates with a community – it is flexible and connects places and people. Djakanimba Pavilions: cheeky, vibrant, harmonious, and as inclusive as the senior man they're named after, Victor Hood. They are adaptable, modular, cool, light, local, theatrical and fun. But it is the space between the pavilions where crowds gather to dance, share, celebrate and recover. As part of the Ghunmarn Cultural Precinct, these prefabricated pavilions bring new possibilities to Wugularr[8] (Beswick) community. Prefab construction used here as a kit-of-parts for the Djakanimba Pavilions is a typological alternative that resolves the most complicated situations and programmes – like fold-up (and fold-down) refuge for inveterate travellers, itinerant exhibition space, training places (Fig. 7.7).

Fig. 7.7 Djakanimba Pavilions during the moon rise, Wugularr, Northern Territory
'Winner of the inaugural Nicholas Murcutt Award for Small Project Architecture, Djakanimba Pavilions took a modest budget and created an adaptable cultural space that is helping to drive regeneration and growth to the local Indigenous community of Beswick. These projects offer a legacy of positive urban spaces, connections and approaches that will influence how their cities function and are perceive' AIA national jury chair, Shelley Penn. (Photo by Peter Eve, courtesy Insideout Architects)

[8] Wugularr is on the banks of the Waterhouse river, approximately 116 km southeast of Katherine in the Northern Territory, and 31 kms from the Barunga Community. The Traditional Owners are the Bagala people.

Fig. 7.8 Jupiter, Djilpin Arts. (Photo by Peter Eve, Courtesy Insideout Architects)
'When we dance and our feet hit the ground, they listen – they go "ohhh that song's still going". it's a deep connection of our families in the land. Not on the land – in the land. It's beautiful isn't it?' (Balang (The late Balang Tom. E. Lewis, Djilpin Arts Cultural Director and Traditional Owner), describing Jupiter)

Walking with Spirits Festival is the showcase event for the Ghunmarn Culture Precinct. Held annually on Jawoyn country, the event maximises local indigenous participation and facilitates opportunities for artistic development and exchange on a local, national and international basis. Djakanimba Pavilions provide flexible kitchen, accommodation, gallery and workshop space. When you visit Djakanimba Pavilions, you support an indigenous enterprise that facilitates remote art, performance, cultural continuity and exchange, training and employment; and perhaps that useful souvenir could be seeing the in-between spaces with new eyes (Fig. 7.8).

They're not unused spaces, they're fluid and used spaces – a renaissance of change, creating places for exchange. They shape our environments into more nurturing, equal and diverse places. You might also see the politics of space and place – who is allowed to use what, and who has the right to what, and what it is that makes somewhere a good place to be. Often unwritten, seeing and respecting these cultural nuances demonstrate respect for other people and their ways which, when respected, lead to acceptance of outsiders into a group. The space between is layered and unusual; and this mix of space, activity, atmosphere, materials, light, warmth, people, shared stories and connection needs deliberate consideration. It enables social cohesion. It's good public space.

Working alongside the Djilpin Arts Crew, outside contractors and local Wugularr builders, Insideout Architects have created an inspiring and joyful setting for the continuing regeneration of the Wugularr community, healing a community and creating a future. As a regional woman in the construction industry, it also makes me happy and inspired to work with other regional North Australian people and their communities to deliver exceptional outcomes. It is also about getting local women involved in construction – the women I've worked with bring a different perspective to the industry – the exchange of experiences, ways of doing things, care and personal or cultural insights and provides tangible improvement of on-site work culture and respect.

7.8 Will Happiness Find Us in This Place?

We transform our lives through planning for a place. Developing clear social objectives to manage and implement change in the places we live is as much a part of the design process as it is the end product. Human interventions build connections, and the intersection between human happiness and urban design is social connection. People who are socially connected are more resilient, and they are more productive at work. Wellbeing is different for everyone and is based on relationships, something that happens with people in place. Some of these relationships are physical and ecological, such as access to healthy food and connection to nature; others are social and psychological, such as being part of a community and having positive emotions. Wellbeing depends as much on what happened yesterday as it does on what is happening in the moment or could happen in the future. We like spaces that flirt with us – the mystery, the disruption of accepted thought – vernacular structures and settlements to create something suited to human needs. Give people the tools to cocreate their own places and they feel a greater sense of pride, so look outwards and drive innovation through problem-finding rather than just problem-solving. Enable local job creation and employment; support business or enterprise development opportunities.

7.9 Conclusion

The design process anchors buildings like *Tjulyuru* to their realities of place and ensures what is built is what was conceived with Ngaanyatjarra people. *Tjulyuru* was a bicultural group process and collaboration that extended beyond the job site and included the whole of the Ngaanyatjarra Lands and its 2000 *Yarnangu*[9]. This inclusive process produced a place that provides *Yarnangu* with an opportunity to promote, share, maintain and continue to grow their culture. The key principle is

[9]The name used by members of the Western Desert block of Aboriginal people to describe themselves. The original meaning of the word was 'human being, person', 'human body' and is rarely or never applied to non-Aboriginal people.

that *Tjulyuru was* designed *with*, not designed *for Yarnangu*. Good architecture should never be about the architect's ego. It's about being 'hands on', working and learning from each other, so in the end you 'just do it!'

Architecture is a bridge between theory and practice, and practitioners need to be conscious of their ability to speak, or at the very least understand, the local language; and use this understanding to explore the relationships between practice and place. So, ask people what kind of community/town/city/society they want – greener, slower, cooperative, fairer … And don't be too driven by cost, because often what costs more initially delivers real long-term savings[10], and a long-term perspective is central to sustainable design. The way we design our city systems, places and spaces has a strong effect on how we regard other people. Conviviality makes people think more altruistically and more collaboratively. Climate change demands ecologically restorative cooperation, and a connected and inclusive society will come up with solutions.

Resilience in design is much more than the buildings, it is about social value. In ancient Greek, the word *krisis* means decision, required when things come to a head and could no longer be avoided. That ancient spirit is what resilience should aim to recover – the most valuable craft of metamorphic reconfiguration. Today, we often hear architects and engineers being challenged to produce work that is sustainable and resilient but, analytically, 'sustainable' and 'resilient' part company in this sort of remediation. You don't want the one party or faction to be sustainable in the sense of enduring and time-proof; you do want the system to be resilient, springing back from inadequacies or incapacities in any one part.

Calmly decide what to do when faced with a crisis – and don't let hysteria or terror drive you. We need to demand from our designers and developers that they take into account the emotional effect of what they do if we are to achieve healthier, prosperous and resilient communities that draw us together and bring out our best. And we need to demand that they design with us. We all need a moral code.

References

Alexander, C., Ishikawa, S., Silverstein, M., Jacobson, M., Fiksdahl-King, I., & Angel, S. (1977). *A pattern language: Towns. Buildings. Construction.* New York: Oxford University Press.

Donovan, J. (2013). *Designing to heal: Planning and urban design response to disaster and conflict.* Collingwood: CSIRO Publishing.

Goad, P. (2005). *Troppo: Architecture for the top end* (2nd ed.). Sydney: Pesaro Publishing.

Litchfield, H. (2012). *Wallaby Man.* (The Tania Dennis collection). Litchfield family collection, Unpublished manuscript.

[10] There are many abandoned or unused buildings that stand as expensive testament to the folly of an externally imposed 'good idea'.

Chapter 8
Health as a Building Block for Resilience

One essential condition for sustainable and resilient communities in northern Australia is that the majority of the population enjoys a reasonable level of physical and mental health, most of the time. Ill health, of body or mind, will limit the capacity of individuals to participate actively in the northern economy or contribute to the everyday community activities that bring a sense of social cohesion. Although there are challenges to both health and health care in northern communities, there are also a number of resilience-enhancing factors in terms of health and wellbeing, most notably connection to land and country, strong social networks and flexibility to innovate to meet need.

There is no doubt that appropriate housing, education, access to healthy food and economic, physical and emotional security are all vital determinants of good health. However, access to appropriate health services, ideally close to home, is important in terms of delivering the required preventive and curative health care to maximise the potential of community members for health and wellbeing. Rural and remote communities remain underserved in terms of access to health care. Yet innovation, born of necessity in the provision of health care to rural and remote residents, has been a stimulus to the evolution of remote health service delivery. Health services across Australia tend to adopt the lessons learnt from the bush to increase access to services and efficiency of service delivery for a wide variety of populations.

This chapter summarises health status and both challenges and protective factors for northern communities. Then it outlines some challenges for health service delivery and the importance of a strong and stable health service as a community hub or focal point for the community. This chapter concludes by outlining a range of strategies currently actively curated by people in isolated communities to best deliver optimal health care and maximise health and wellbeing.

Sarah Larkins is the primary author of this chapter.
Sarah Larkins and Keith Noble have contributed more to this chapter.

K. Noble et al., *Agriculture and Resilience in Australia's North*,
https://doi.org/10.1007/978-981-13-8355-7_8

8.1 Health Status: Threats and Protective Factors

For many years, a deficit discourse about health and health care has been prominent in government reports, academic literature and the media. In the Australian setting, this focuses on the well documented lower life expectancy, and higher rates of particular diseases, particularly cardiovascular disease, cancer and mental health conditions amongst rural and remote populations, here defined as all areas outside of Australia's major cities (Australian Institute of health and Welfare 2017). Higher rates of risk factors for cancer and chronic disease, such as obesity, cigarette smoking and heavy alcohol use are also recorded in the bush (Australian Institute of Health and Welfare 2016, 2017). These perhaps reflect the lower average socio-economic status of people residing in northern Australia and the lack of reach of health promotion campaigns designed and largely promulgated in southern urban centres. Lower education levels and higher Aboriginal and Torres Strait Islander populations in northern Australia (both groups known to have higher rates of lifestyle risk factors and chronic disease) as well as long travel distances in remote areas contribute to these disease gradients. Occupational risks are also an issue, with high rates of mental health issues amongst fly-in, fly-out (FIFO) and drive-in, drive-out workers, mostly in the mining industry (Torkington et al., 2011) and climate risks including sun exposure playing a role.

What is less well understood is the degree to which rurality in and of itself contributes to these poor health outcomes, particularly if the contribution of reduced access to the full range of primary health care (including health promotion and prevention) and secondary and tertiary health services is taken into account. Those that have attempted to address this have found a complicated picture, with individual factors (stoicism and reluctance to seek help, together with lifestyle risks), socio-economic factors and accessibility factors all playing a major role (Australian Bureau of Statistics 2017; Duckett 2016). It is clear that there is place-based inequality, with communities such as Mt. Isa and Palm Island in Queensland having hospital admission rates that are potentially preventable with good primary care at least 50% above the state average for the last 10 years (Duckett 2016).

The stoicism of rural residents, particularly farmers, is well documented and a point of pride among many rural residents (Buikstra et al. 2010). This traditional rural stoicism, particularly prevalent among men, might be adaptive in allowing residents to 'roll with the punches' of life on the land and bounce back from climatic extremes and natural disasters (Morrissey and Reser 2007). However, this 'hard masculinity' also increases the risk of delayed presentation and diagnosis of potentially treatable health conditions. In addition, stoicism compels the bottling up of anxieties and distress – hence increasing the risk of serious mental health issues including major depression and culminating in the worst-case scenario in suicide (Wainer and Chesters 2000). Unfortunately, due to access to more lethal means such as firearms, suicide attempts are more often 'successful' in rural and remote areas.

A realistic appraisal of the mental health impact of life in the rural and remote north, avoiding both 'romanticism and despair', is essential for communities to address these difficult realities (Wainer and Chesters 2000), and many are indeed attempting to do so through the provision of men's sheds and similar initiatives. Health care-seeking behaviour differs with the provision of acceptable and accessible services close to home – the challenge is for the health system to adapt to provide this in sustainable and effective ways.

8.1.1 Land and Belonging as Strength

To counter some of these discouraging indicators, it is clear that living in rural and remote areas brings a number of resilience-enhancing factors, critically important for both individual and collective health and wellbeing (Wainer and Chesters 2000). The first of these is connection to country or land and a sense of deep belonging to a place and the natural environment and of being 'at home' there and of this enhancing wellbeing. This is considered most frequently and explicitly in the context of the health and wellbeing of Aboriginal peoples and Torres Strait Islanders, where indeed direct links between relationship with country and morbidity and mortality have been demonstrated (Burgess et al. 2009). Burgess and colleagues (Burgess et al. 2008, 2009) define relationship with country as encompassing an interdependent relationship between Indigenous peoples and their ancestral lands and seas. For example, among a group of Yolngu people from a community and surrounding homelands in Arnhem Land, the degree of self-rated involvement in 'caring for country' activities was correlated with a range of positive health indicators (including activity levels, better diet, lower obesity and less diabetes, among others; Burgess et al. 2009).

More recently recognised in academia (although long recognised by farmers and other rural-dwellers themselves) is a related sense of belonging to country and how this sense of place and belonging, often transgenerational, can strengthen wellbeing (Buikstra et al. 2010; Hegney et al. 2007). So connection to land and country is vitally important to wellbeing, but so too are connections between people in communities of place (Taylor et al. 2008). Social connectedness or cohesion fulfils essential social needs and enhances health and wellbeing. In contrast to urban centres, where the tendency is to cocoon within a nuclear family, residents of small rural towns often have extremely strong social networks and routines. This is facilitated by living in a town of a size where people know everyone and run into each other (Farmer et al. 2016). While those living in isolated remote stations face more geographical challenges, these families have often developed creative solutions to overcome their isolation and will travel large distances for regional events in order to foster and maintain this sense of community (Fig. 8.1).

Fig. 8.1 The annual Alice Springs show attracts crowds of all ages and ethnicity from three adjoining states – the Northern Territory, South Australia and Western Australia. (Photo by Tania Dennis)

8.2 Challenges in Health Service Delivery to Remote Areas

The harsh and vast terrain of northern Australia provides challenges in service delivery, not least health service delivery. Dispersed small populations over enormous and often inhospitable geographical areas preclude economies of scale and mean that often residents need to travel large distances to access primary health-care services, let alone more specialised services. Recent reports from the Grattan Institute, among others, have highlighted the degree to which current policy settings are failing to deliver an efficient and effective primary care-based health systems to northern Australia and other rural and remote settings (Duckett 2016; Swerissen and Duckett 2018). Furthermore, over the last 20 years, policy changes at jurisdictional level, together with the drift away from generalism among health-care providers, have reduced the range of procedural services available in small rural settings. This means that rural and remote residents need to travel vast distances to access health services. A good example of this is birthing services; in Queensland, the number of public birthing units dropped by 43% due to closures in the 10 years between 1995 and 2005, which had the effect of transferring both risk and cost from the providers of the services to individual consumers of these services (Evans et al. 2011).

Local councils and regional health networks (known as primary health and local hospital networks under current policies) face the challenge of ensuring that the needs of their communities are met through health services delivering accessible high-quality care (in this context, quality includes elements of clinical effectiveness, safety and the patient experience). Accessibility of health services is again the subject of much debate but encompasses geographical access, opening hours, affordability, cultural appropriateness and acceptability by the population being served, together with an awareness of the service (Saurman 2015).

Best practice in this area recognises that citizens resident in all parts of the country should have access to an 'essential basket of services' comprising a suite of necessary primary health-care services and access to specialised services as needed. This usually involves management of minor acute conditions, ongoing management of chronic conditions, basic sexual and reproductive health services, child health, mental health care, health promotion and prevention services, emergency retrieval services when required and access to specialised services when needed. The usual model of private general practice that is dominant in southern Australia and major regional centres usually proves to be financially non-sustainable in all but the larger rural towns, so alternatives are used. There are a variety of models used to deliver this basket of services in rural and remote Australia (Wakerman et al. 2008). These include (i) Aboriginal community-controlled health services (ACCHS; single service or hub and spoke models), (ii) primary care clinics staffed by remote area nurses and Aboriginal health practitioners with visiting doctors, (iii) fly-in, fly out services such as those operated by the Royal Flying Doctor Service, and (iv) rural generalist models (medical superintendent of a small rural hospital with the right to private practice).

8.2.1 Health Workforce

The biggest ongoing problem in health service provision across northern Australia relates to ongoing shortages and maldistribution in the health workforce (Department of Health 2016b; Swerissen and Duckett 2018). Overall, in Australia, we have an ample supply of doctors; however, they continue to be located largely in capital cities and urban centres, with a skewed balance of specialist to generalist practitioners that current training figures suggest is likely to worsen (Department of Health 2016a). There is also a shortage of nurses and midwives, along with (concerning) ageing of this workforce. The status quo in medical selection and training fuels this maldistribution, whereby the majority of health professional training takes place in southern metropolitan centres, providing training for the children of the urban elite, who unsurprisingly go on to work as subspecialists in urban areas (Murray et al. 2012). Perverse financial incentives for subspecialisation exacerbate this issue, despite the priority for a generalist workforce with a broad range of skills living and practising where there is the greatest need (i.e. Northern Australia).

Other contributors to the rural health workforce shortage are the perception of limited opportunities for partner employment and for children's schooling (particularly in secondary school years), so rural health workforce agencies invest considerable time and effort into matching practitioners and their families with communities. Limited workforce retention and high turnover raise the issue of continuity of care –

practitioner churn provides suboptimal care to rural and remote populations and leaves local health services with the burden of huge locum fees to fill gaps in service delivery. Conversely, there is a group of health professionals whose personal characteristics, particularly when supported by appropriate infrastructure and a regional network of supportive health professionals and referral pathways, are well-equipped for rural and remote practice (Matheson et al. 2016). The recently developed Rural Generalist Pathway, developed and successfully operationalised in Queensland and now expanding nationally, is a good example of the evolution of postgraduate training pathways in response to need (Sen Gupta et al. 2013). This model is being adapted for use in Pacific neighbour countries such as Papua New Guinea and Japan.

8.2.2 *Training a Fit-For-Purpose Health Workforce*

In both Australia and north America, evidence for the effectiveness of the so-called rural pipeline is evolving, yet it continues to be the exception rather than the rule. This pipeline involves (i) recruiting students from rural and remote areas, (ii) training them in regional and rural areas with a primary care-focused curriculum orientated to local priorities and (iii) providing postgraduate pathways and training programs that do not require relocation to urban areas at a stage of life where people are forming long-term partnerships (Murray and Wronski 2005). Internationally, some schools of health professional education orient themselves to training a 'fit-for-purpose' health workforce, that is, a health workforce equipped with the knowledge, attitudes and skills to meet the health needs of the communities they serve, with a focus on underserved populations. One community of practice of 12 such schools is the Training for Health Equity Network (THEnet). These schools, located in diverse low- and high-income settings, have collectively developed a framework for critically reflecting on progress against their social missions (Larkins et al. 2013). They are working to follow graduate outcomes to assess the extent to which they are practising in areas of need (Larkins et al. 2015).

The James Cook University (JCU) medical school, a founding member of THEnet, was established in 2000 as the first new medical school in Australia located out of a capital city. The JCU school has an explicit social mission to train doctors to meet the health needs of rural, remote, Indigenous and tropical populations. To support this mission, selection processes are orientated towards rural, regional and Aboriginal and Torres Strait Islander applicants. All training takes place in the north, all students have at least 20 weeks of small rural and remote placements and postgraduate training pathways have been developed in a variety of generalist disciplines. Outcomes have been very positive, with 12 cohorts now graduating and practising largely in generalist disciplines in areas of need (Woolley et al. 2018). This has had spin-off benefits, including the development of thriving stable service and training centres out of previously fragile health services, such as those in Cloncurry, Longreach, Cooktown and Proserpine (Sen Gupta et al. 2009).

Similarly, training for nurses, dentists, pharmacists and allied health professionals requires decentralised training and support to ensure a sustainable health workforce. A particular priority issue in the north is the need for better training pathways, support and career structures for Aboriginal and Torres Strait Islander health practitioners (Hill et al. 2018). These workers play a vital role in strengthening health systems, especially in the remote north. In addition to their direct clinical role, their importance as cultural brokers and links with the community are vital to successful operation of a health service, and as local residents they are often a beacon of stability in a transient workforce.

8.3 The Health Service as a Barometer

In many small rural and remote communities, the maintenance of a health service is an indicator of community viability and sustainability. Health services are often significant employers and indirectly generate work for a variety of other services in the town. Like the loss of a local school, loss of a local health service can be a major blow, and a significant event in the decline of a small community as the population, particularly young families and the elderly (more frequent health service users), then tends to relocate to where health services are more accessible. Many rural councils recognise the importance of maintaining a viable service and invest in a variety of strategies to ensure the sustainability of the health service. These might include a council investment in infrastructure and practice support, enabling a practitioner to provide several years of service without committing to a major financial investment, or the provision of staff housing, jobs for partners or other tangible support. These "walk-in, walk-out" models, such as that operated by Rural and Remote Medical Services Ltd. (RaRMS) in NSW, remove some of the bureaucratic barriers for health professionals in committing to rural work.

Councils and communities alike are pragmatic in accepting changing realities in terms of the rural professional workforce. A lifetime commitment to one rural community is now uncommon; however, many young skilled health professionals are prepared to provide a commitment of 5–10 years' service – a very helpful contribution in areas of scarcity. As active participants in attracting and supporting their own health workforce, there is much that rural community members through their local councils and service groups can (and do) arrange. Welcomes and support provided to incoming health practitioners are legion. This support extends to visiting health professional students, who are usually warmly welcomed, with the hope that a proportion of them might come back and practise in future years.

Vignette 8.1: Flinders Medical Group: Sustainability in Health Service Provision

Cloncurry is a town of approximately 3000 residents in outback Queensland, 9 h' drive west of Townsville or 1 h east of Mt. Isa. Traditionally, Cloncurry has been a town that struggled to attract health professionals. In 2004, Dr. Sheilagh Cronin started providing locum relief in the area and soon after, together with Dr. Bryan Connor, decided to build a teaching-focused medical practice to sustainably meet the needs of the region and train future health professionals. They constructed a six-consulting room practice and built it up as a strong teaching practice.

Subsequently two sister practices in the Sunshine Coast Hinterland have been purchased. These play dual roles: a source of qualified, senior doctors to relieve the Cloncurry doctors when necessary and a practice location for senior Cloncurry doctors to go for respite from the intensive practice in the remote area. The practice has grown and has also become a centre for nursing and Intensive Health Worker training. Recently a graduate from JCU medical school and now a rural generalist doctor has taken on a senior role in Cloncurry.

8.4 Innovation in Service Delivery to Meet Need

8.4.1 Codesign of Service Solutions

Place-based, needs-based health service planning is an essential element in the improvement and design of health services for rural and remote communities and should ideally be combined with planning for other service delivery in the community-of-place. When asked to contribute to needs-based health service planning, rural community members are realistic and practical about the provision of services in a particular community and options for improving the accessibility of other services (Farmer and Nimegeer 2014). Community participation in health service design has now become theoretically mainstream and recognised in health systems quality frameworks and standards (Australian Commission on Safety and Quality in Health Care 2012). However, all too often, operationalisation of community participation is little more than notices in waiting rooms inviting feedback or suggestions, rather than meaningful participatory planning processes.

Acknowledging the critical role of health-care providers who are long-term residents in a community as "boundary crossers" (Farmer and Kilpatrick 2009; Kilpatrick et al. 2009), planning services with grass-roots community members, health-care providers and stakeholders and configuring these in the most efficient way irrespective of funding silos can deliver real improvements in health-care access and effectiveness. Primary Health Networks and Local Hospital Networks can have a critical role in facilitating these collaborative planning processes, and both community groups or councils or academic partners can provide useful guidance through the process. The role of facilitation itself in steering or shaping the process is very important in influencing the degree of participation (Carlisle et al. 2018; Harvey and Lynch 2017).

8.4.2 *Hub and Spoke Model of Service Delivery*

Throughout remote Australia, green medical chests provided through the Royal Flying Doctors' Service (RFDS) are ubiquitous on the vast stations otherwise isolated from health care. These chests, together with the accompanying telephone or Voice over Internet Protocol (VOIP) support from remote doctors or nurses, are a critical link to emergency health care for isolated station residents (both permanent residents and contract workers).

Similarly, many remote residents and rural residents access primary health clinics staffed by remote area nurses and sometimes also Aboriginal or Torres Strait Islander Health Workers or Practitioners (hereafter called Indigenous Health Workers; IHW). In some cases, these services are supported by fly-in medical staff who visit for clinics on a regular basis (funded through either the RFDS or government) to deliver complementary services.

8.4.3 *Evolving Workforce Models: The Role of Delegated Practice Models and Task Substitution*

In areas of workforce shortage, a multi-skilled and flexible generalist workforce is very important. Delegated practice models where primary health-care practitioners practice under the supervision and delegated authority of a doctor or remote area nurse (usually but not always co-located with them) can very usefully supplement the capacity of a rural or remote health service and increase the sustainability of the posting through enabling sharing of the often onerous on-call responsibilities. Indigenous Health Workers have long filled this role and can build up experience and training to fulfil a variety of essential roles within the health system. Physician assistants are mid-level health-care providers most commonly found in north America. More recently, physician assistants have been utilised in the Australian health-care system, where they are perceived as very useful in rural areas, yet their status is still a little uncertain due to delays with registration status. Mid-level providers can develop a range of special skills related or complementary to the scope of practice of their supervisory doctor and can provide sustainability, particularly in towns where there is too much workload for one doctor but not enough for two (or the community is unable to attract two). Importantly, people in rural and remote communities want a stable health-care provider who knows them and knows their town and are less concerned about the professional background of the health-care provider (Crossland 2011).

8.5 Technological Solutions for Health in Isolated Communities

As our technological capacity increases, telemedicine solutions are emerging as critical tools in ameliorating isolation from rural and remote residence. The Internet of things (IoT) is opening a world of possibilities in terms of connectivity for health and health care. These technological solutions are in a phase of rapid development and relatively rapid uptake where the need is the greatest, although in the most remote areas Internet access and speeds still limit their utility. Like many needs-based innovations, in many cases the speed of implementation is outpacing the evaluation science in terms of models and their effectiveness, so published evidence is sparse.

Broadly, these technological solutions fall into four groups: (i) increased access to specialised services not otherwise available locally; (ii) personal support for remote health-care providers through e-health guidance, remote supervision or access to opinions; (iii) systems support through monitoring quality of care or population health; or (iv) remote monitoring of the health status of individuals through various e-health applications (including the provision of tailored health promotion and advice). All of these options, of course, exist in a world of ever-increasing health resources available to individuals through the World Wide Web. Increasing proficiency in information technology and computer and Internet access at population level is allowing greater usage, although access, health and information technology literacy are still unevenly distributed. Each of these four groups of solutions will now be discussed.

8.5.1 Increased Access to Specialist Services Closer to Home

Enabling access to specialist health services closer to home has long been an aspiration for rural residents, where the direct and indirect out-of-pocket costs of attending regional centres for specialist services, for routine monitoring or for follow-up visits can be prohibitive. In the past, the only alternative to patient travel was specialist outreach services, where a specialist practitioner from a capital city or regional centre makes regular (if infrequent) visits to rural centres to deliver specialist services (Gruen et al. 2002). This model still has a place but can also be inefficient if the booking, planning and travel arrangements for these clinics are not well coordinated with the local primary health-care team, and tends to operate in parallel with, rather than in harmony with the primary care team, with limited capacity strengthening intent.

An alternative is using telemedicine linkages to link rural and remote residents, sitting close to home with a trusted local primary care provider, with specialist practitioners in regional centres. Across North Queensland this has been done extensively for tele-oncology services (Sabesan et al. 2012a), used for both monitoring and follow-up of people with cancer and for remote supervision of chemotherapy for selected patients and conditions. These models have been evaluated and found to be acceptable for both non-Indigenous (Sabesan et al. 2012b) and Indigenous patients (Mooi et al. 2012), safe (Chan et al. 2015) and cost-effective (Thaker et al. 2013), and these results have led to broader uptake across the region.

Other well-known models of telemedicine include remote ophthalmology support for retinal photographs for people with diabetes taken by local practitioners (Glasson et al. 2016), and the telederm program, whereby local practitioners can send photos of skin lesions and rashes to a remote dermatologist (http://www.ehealth.acrrm.org.au/provider/tele-derm). Similar programs are available for remote cardiology and orthopaedics opinions based on electronic transmission of electrocardiograms (ECGs) and X-rays or scans. These systems are similarly useful for allied health practitioners – for example, the use of tele-rehabilitation to allow rural or remote residents to continue to receive support from physiotherapists or occupational therapists after they continue their recovery from a stroke or other neurological condition (Kairy et al. 2009), although little has been published about this in the Northern Australian setting.

8.5.2 Personal Support for Remote Health-Care Providers

In an area of scarcity, enabling every member of the health workforce to practise at the upper (optimal) end of their scope of practice is vital. This needs to occur with support from more specialised or experienced practitioners, but this support can be provided either on site or remotely, with support via telemedicine or telephone for difficult clinical problems. The Remote Vocational Training Scheme has utilised remote supervision, webinars and other means of remote support to allow rural and remote practitioners to undergo vocational training without moving away from their post in an area of workforce need (Wearne et al. 2010). Tele-supervision of trainees in other disciplines has also been found useful, although it is necessary to be mindful of factors enhancing its effectiveness (Martin et al. 2018).

Similarly, remote area nurses provide a range of care, often complex, in extremely challenging environments. Professional and personal support services are provided through the Council of Remote Area Nurses of Australia (CRANA) through a variety of means, including the well-used Bush Crisis Line. This service, initially provided by telephone and more recently moving to Internet-based services, has proved time and again to be of great value in supporting remote area health professionals as they deal with the traumatic experiences that are often part of this work (Ellis and Kelly 2005).

The last vital area of support for remote health-care providers consists of support by emergency care specialists in regional centres in real time during an acute situation while managing an emergency. This serves two functions – providing specialist level support to remote providers to perform life-saving procedures and best practice care and making appropriate and timely decisions about the need for retrieval to specialist care and the best mechanism for this (Mathews et al. 2008; Robb et al. 2009)

8.5.3 Systems Support and Quality Monitoring

Providing point-of-care decision support tools and management or referral pathways to isolated health-care providers is a further role for technology in optimising health care. This can be done through the Internet or alternatively via mobile phones or other palm-held technology. Through linking with local or regional evidence-based best practice guidelines, a remote practitioner can instantly link with point-of-care clinical decision pathways and algorithms, simplifying the clinical decision-making process.

One of the issues limiting quality improvement in primary care on a broad scale has been the absence of a universal health-care record with a unique identifier for everyone. This is slowly becoming a reality through the integrated electronic medical record (IEMR), although technical issues and privacy concerns have plagued the implementation process. Despite this, as most health services now have local electronic medical records of some kind, it is now possible to automatically extract, aggregate and analyse quality of care against guidelines at service or regional level. This strengthens quality improvement efforts and allows health-care providers to target health care and improvements towards those at greatest need.

8.5.4 Remote Monitoring of the Health Status of Individuals

Empowerment and engagement of individuals in the self-management of their own chronic conditions are important elements of quality care. This is perhaps particularly so in the bush, where access to qualified health-care providers might be at a distance and thus delayed. A promising advance is the use of various devices to assist people to monitor their own health conditions in their own home, with electronic transmission (either automatically through sensors or via upload at agreed intervals) to a specialist chronic disease nurse or other practitioner located at a distance in a rural town or larger centre. Through this, it is possible to monitor the stability (or otherwise) of the person and their condition and either predict the need for access to care or hospitalisation or, ideally, make changes to avert this deterioration and avoid the need for hospital level care. Results at this stage remain variable, but there is much excitement and promise in the approach, and it is valued and accepted by rural and regional residents (Carlisle and Warren 2013).

8.6 Conclusion

Although residents of rural and remote parts of Australia continue to face a greater burden of disease and limited access to health care when compared with their city-dwelling sisters and brothers, there is a considerable cause for optimism. Rural and regional communities have a variety of resilience enhancing characteristics and continue to innovate in terms of health workforce and health service design to meet the needs of their populations. At the forefront of the use of technology in health care, there is now hope that the tyranny of distance needs no longer be a barrier to top quality health care and healthy resilient individuals and communities in the bush.

References

Australian Bureau of Statistics. (2017). *Health service usage and health related actions, Australia 2014–15*. (ABS cat. no. 4364.0.55.002.). Canberra: ABS.

Australian Commission on Safety and Quality in Health Care. (2012). *National Safety and Quality Health Service Standards*. Retrieved from Canberra: https://www.safetyandquality.gov.au/wp-content/uploads/2011/09/NSQHS-Standards-Sept-2012.pdf.

Australian Institute of Health and Welfare. (2016). *Australia's health 2016. Australia's health no. 15*. (Cat. no. AUS 199). Canberra: AIHW.

Australian Institute of health and Welfare. (2017). *Rural and remote health web report. Version 3*. Retrieved from https://www.aihw.gov.au/reports/rural-health/rural-remote-health/contents/rural-health.

Buikstra, E., Ross, H., King, C. A., Baker, P. G., Hegney, D., McLachlan, K., & Rogers-Clark, C. (2010). The components of resilience—perceptions of an Australian rural community. *Journal of Community Psychology, 38*(8), 975–991.

Burgess, C. P., Berry, H. L., Gunthorpe, B., & Bailie, R. S. (2008). Development and preliminary validation of the "Caring for Country" questionnaire: Measurement of an indigenous Australian health determinant. *International Journal of Equity in Health, 7*(26), 1–14.

Burgess, C.P., Johnston, F.H., Berry, H.L., McDonnell, J., Yibarbuk, D., Gunabarra, C., et al (2009). Healthy country, healthy people: The relationship between indigenous health status and "caring for country". *Medical Journal of Australia, 190*(10), 567–572.

Carlisle, K., & Warren, R. (2013). A qualitative case study of telehealth for in-home monitoring to support the management of type 2 diabetes. *Journal of Telemedicine and Telecare, 19*(7), 372–375.

Carlisle, K., Farmer, J., Taylor, J., Larkins, S., & Evans, R. (2018). Evaluating community participation: A comparison of participatory approaches in the planning and implementation of new primary health-care services in northern Australia. *International Journal of Health Planning and Management, 33*(3), 704–722. https://doi.org/10.1002/hpm.2523.

Chan, B. A., Larkins, S. L., Evans, R., Watt, K., & Sabesan, S. (2015). Do teleoncology models of care enable safe delivery of chemotherapy in rural towns? *The Medical Journal of Australia, 203*(10), 406.

Crossland, L.J. (2011). *Perceptions of the roles and skills of primary health care professionals: Implications for innovative and sustainable rural primary health care delivery*. James Cook University.

Department of Health. (2016a). *Medical training review panel. 19th report*. Canberra: AGPS. Retrieved from http://www.health.gov.au/internet/main/publishing.nsf/content/8795A75044FBB48CCA257F630070C2EE/$File/Medical%20Training%20Review%20Panel%20nineteenth%20report.pdf.

Department of Health. (2016b). *Medical workforce factsheet 2016*. http://hwd.health.gov.au/webapi/customer/documents/factsheets/2016/Medical%20workforce%20factsheet%202016.pdf.

Duckett, S. (2016). *Perils of place: Identifying hotspots of health inequality*. Melbourne: Grattan Institute.

Ellis, I., & Kelly, K. (2005). Health infrastructure in very remote areas: An analysis of the CRANA bush crisis line database. *Australian Journal of Rural Health, 13*(1), 1–2.
Evans, R., Veitch, C., Hays, R., Clark, M., & Larkins, S. (2011). Rural maternity care and health policy: Parents' experiences. *Australian Journal of Rural Health, 19*(6), 306–311. https://doi.org/10.1111/j.1440-1584.2011.01230.x.
Farmer, J., & Kilpatrick, S. (2009). Are rural health professionals also social entrepreneurs? *Social Science and Medicine, 69*(11), 1651–1658.
Farmer, J., & Nimegeer, A. (2014). Community participation to design rural primary healthcare services. *BMC Health Services Research, 14*(1), 130. https://doi.org/10.1186/1472-6963-14-130.
Farmer, J., De Cotta, T., McKinnon, K., Barraket, J., Munoz, S.-A., Douglas, H., & Roy, M. J. (2016). Social enterprise and wellbeing in community life. *Social Enterprise Journal, 12*(2), 235–254.
Glasson, N.M., Crossland, L.J., Larkins, S.L. (2016). An innovative Australian outreach model of diabetic retinopathy screening in remote communities. *Journal of diabetes research, 2016.*
Gruen, R. L., Weeramanthri, T., & Bailie, R. (2002). Outreach and improved access to specialist services for indigenous people in remote Australia: The requirements for sustainability. *Journal of Epidemiology & Community Health, 56*(7), 517–521.
Harvey, G., & Lynch, E. (2017). Enabling continuous quality improvement in practice: The role and contribution of facilitation. *Frontiers in Public Health, 5*, 27. https://doi.org/10.3389/fpubh.2017.00027.
Hegney, D.G., Buikstra, E., Baker, P., Rogers-Clark, C., Pearce, S., Ross, H., et al (2007). Individual resilience in rural people: A Queensland study, Australia. *Rural and Remote Health, 7*(14), 1–13.
Hill, K., Harvey, N., Felton-Busch, C., Hoskins, J., Rasalam, R., Malouf, P., & Knight, S. (2018). The road to registration: Aboriginal and Torres strait islander health practitioner training in North Queensland. *Rural and Remote Health, 18*, 3899. https://doi.org/10.22605/RRH3899.
Kairy, D., Lehoux, P., Vincent, C., & Visintin, M. (2009). A systematic review of clinical outcomes, clinical process, healthcare utilization and costs associated with telerehabilitation. *Disability and Rehabilitation, 31*(6), 427–447.
Kilpatrick, S., Cheers, B., Gilles, M., & Taylor, J. (2009). Boundary crossers, communities and health: Exploring the role of rural health professionals. *Health and Place, 15*(1), 284–290. https://doi.org/10.1016/j.healthplace.2008.05.008.
Larkins, S., Preston, R., Matte, M.C., Lindemann, I.C., Samson, R., Tandinco, F.D., et al (2013). Measuring social accountability in health professional education: Development and international pilot testing of an evaluation framework. *Medical Teacher, 35*(1), 32–45. https://doi.org/10.3109/0142159X.2012.731106.
Larkins, S., Michielsen, K., Iputo, J., Elsanousi, S., Mammen, M., Graves, L., et al (2015). Impact of selection strategies on representation of underserved populations and intention to practise: International findings. *Medical Education, 49*(1), 60–72. https://doi.org/10.1111/medu.12518.
Martin, P., Lizarondo, L., & Kumar, S. (2018). A systematic review of the factors that influence the quality and effectiveness of telesupervision for health professionals. *Journal of Telemedicine and Telecare, 24*(4), 271–281.
Matheson, C., Robertson, H., Elliott, A., Iversen, L., & Murchie, P. (2016). Resilience of primary healthcare professionals working in challenging environments: A focus group study. *British Journal of General Practice*. https://doi.org/10.3399/bjgp.16X685285.
Mathews, K. A., Elcock, M. S., & Furyk, J. S. (2008). The use of telemedicine to aid in assessing patients prior to aeromedical retrieval to a tertiary referral centre. *Journal of Telemedicine and Telecare, 14*(6), 309–314.
Mooi, J. K., Whop, L. J., Valery, P. C., & Sabesan, S. S. (2012). Teleoncology for indigenous patients: The responses of patients and health workers. *Australian Journal of Rural Health, 20*(5), 265–269.
Morrissey, S. A., & Reser, J. P. (2007). Natural disasters, climate change and mental health considerations for rural Australia. *Australian Journal of Rural Health, 15*(2), 120–125.

Murray, R., & Wronski, I. (2005). When the tide goes out: Health workforce in rural, remote and indigenous communities. *Medical Journal of Australia, 185*, 37–38.

Murray, R., Larkins, S., Russell, H., Ewen, S., & Prideaux, D. (2012). Medical schools as agents of change: Socially accountable medical education. *Medical Journal of Australia, 196*, 1–5.

Robb, M., Close, B., Furyk, J., & Aitken, P. (2009). Emergency department implications of the TASER. *Emergency Medicine Australasia, 21*(4), 250–258.

Sabesan, S., Larkins, S., Evans, R., Varma, S., Andrews, A., Beuttner, P., et al (2012a). Telemedicine for rural cancer care in North Queensland: Bringing cancer care home. *Australian Journal of Rural Health, 20*(5), 259–264. https://doi.org/10.1111/j.1440-1584.2012.01299.x.

Sabesan, S., Simcox, K., & Marr, I. (2012b). Medical oncology clinics through videoconferencing: An acceptable telehealth model for rural patients and health workers. *Internal Medicine Journal, 42*(7), 780–785.

Saurman, E. (2015). Improving access: Modifying Penchansky and Thomas's theory of access. *Journal of Health Services Research and Policy*. https://doi.org/10.1177/1355819615600001.

Sen Gupta, T., Murray, R., Beaton, N., Farlow, D., Jukka, C., & Coventry, N. (2009). A tale of three hospitals: Solving learning and workforce needs together. *Medical Journal of Australia, 191*(2), 105–109.

Sen Gupta, T., Manahan, D., Lennox, D., & Taylor, N. (2013). The Queensland health rural generalist pathway: Providing a medical workforce for the bush. *Rural and Remote Health, 13*, 2319.

Swerissen, H., & Duckett, S. (2018). *Mapping primary care in Australia*. Retrieved from Melbourne.

Taylor, J., Wilkinson, D., & Cheers, B. (2008). *Working with communities in health and human services*. Melbourne: Oxford University Press.

Thaker, D. A., Monypenny, R., Olver, I., & Sabesan, S. (2013). Cost savings from a telemedicine model of care in northern Queensland, Australia. *The Medical Journal of Australia, 199*(6), 414–417.

Torkington, A., Larkins, S., & Sen Gupta, T. (2011). The psychosocial impacts of fly-in fly-out and drive-in drive-out mining on mining employees: A qualitative study. *Australian Journal of Rural Health, 19*(3), 135–141. https://doi.org/10.1111/j.1440-1584.2011.01205.x.

Wainer, J., & Chesters, J. (2000). Rural mental health: Neither romanticism nor despair. *Australian Journal of Rural Health, 8*(3), 141–147.

Wakerman, J., Humphreys, J. S., Wells, R., Kuipers, P., Entwistle, P., & Jones, J. (2008). Primary health care delivery models in rural and remote Australia – a systematic review. *BMC Health Services Research, 8*, 276.

Wearne, S., Giddings, P., McLaren, J., & Gargan, C. (2010). Where are they now?: The career paths of the remote vocational training scheme registrars. *Australian Family Physician, 39*(1/2), 53.

Woolley, T., Sen Gupta, T., & Larkins, S. (2018). Work settings of the first seven cohorts of James Cook University bachelor of medicine, bachelor of surgery graduates: Meeting a social accountability mandate through contribution to the public sector and indigenous health services. *Australian Journal of Rural Health*. https://doi.org/10.1111/ajr.12418.

Part III
Discussion and Conclusions

For when the One Great Scorer comes
to mark against your name,
he writes - not that you won or lost -
but HOW you played the Game
Grantland Rice

Introduction

It's a familiar idiom, the importance of how you play the game; and generally, a reliable yardstick for maintaining individual integrity and social standing, as society tends to look down on poor losers and boastful winners (notwithstanding the rewards to high-profile sports stars who sometimes don't live up to the rule). But often society's winners and losers are a consequence of policy decisions regardless of how they individually played the game, and in these situations winning or losing is significant. While how we play/interact with others is unquestionably important, the further development of Northern Australia is more than cricket, and the final results of decisions made will matter.

Is there opportunity for an understanding of individual resilience strategies to inform and improve policy and planning processes and decisions in this twenty-first century push for northern development?

Perhaps this is not the right question. Perhaps a more appropriate question is: How can an understanding of individual resilience strategies assist Australians in the development of a shared and agreed vision for Northern Australia, which will then enable policies and planning processes that can work towards achieving this vision?

In this final part, the lessons of history and understanding of Northern Australia provided in Part I are combined with the lived experience of Part II to inform the discussion of how we might better approach this question. Chapter 9 suggests a unifying thread – Fairness, while Chap. 10 explores how this might work. The intention is to prompt thoughts and discussions that in turn, may enable insights and new ideas. Though it would have been lovely to achieve, there is no epiphany here.

Chapter 9
Bringing It Together: The Thread of Fairness

Policy is solid policy if it's owned and backed by the community.
Scullion (2016)

Disasters happen, and it is not only natural events that cause them; they can also be the product of social, political and economic environments. Wisner et al. (2003, p. 4) warn against treating natural disasters as something peculiar or as events that deserve their own special focus, as this risks 'separating "natural" disasters from the social frameworks that influence how hazards affect people, thereby putting too much emphasis on the natural hazards themselves, and not nearly enough on the surrounding social environment'.

As the interviews with study participants were read, and reread, two things became apparent: first, while participants were prepared to talk about their personal disaster experience, no one dwelt on the subject, and most were keen to progress the discussion to other topics; and second, almost all interview participants no matter what their individual circumstances declared at some point that there were other people *worse off* than themselves. At first, a number of explanations were considered responsible:

- A mechanism to avoid being someone complaining about their circumstances.
- A public show of stoicism – *keeping a stiff upper lip/we're alright, we'll manage*.
- Perhaps at the cognitive level, people were setting the event into a broader perspective in which their position did not seem so terrible after all – a transformational coping style often associated with *hardiness* as a personality style (Maddi 1999).

But with time and repetition, particularly when it was obvious that the circumstances of some were especially grim and that hard (possibly irreversible) decisions

Keith Noble has contributed more to this chapter.

K. Noble et al., *Agriculture and Resilience in Australia's North*,
https://doi.org/10.1007/978-981-13-8355-7_9

were being made, it became apparent that this statement about others being *worse off* represented something more. It seemed that this statement related strongly to the participant's personal sense of agency and their perception of how the world operates, specifically:

- Life can be hard – there are no guarantees of success.
- Ultimately, it is up to the individual to find their own way through.
- Feeling sorry for yourself is not going to change things.
- This should not preclude one from feeling compassion for and helping others.

Though no one used the actual word, there was an expression of the *right* way to do things or, to be more precise, the *fair* way. When people feel that they are being treated fairly, that their uniqueness is understood, that they have a sense of belonging and place in the world and that their voice will be a part of the decision-making, 'then they will feel included' (Dillon and Bourke 2016, p. 7). When participants discussed the impact of natural disasters, these were often regarded as an inevitable (if unwelcome) part of life that had to be dealt with; but when talking about human-induced (and possibly preventable) disasters, fundamentally they were saying that their impact was not *fair* and as a consequence they felt more removed from the broader community than just geographically. The important part appeared to be about perceived moral capital, that is, not so much 'what should be done, but how ought decisions be made' (Brown 2016, p. 112). Renn's (2015) work on risk governance provided clarity on this concept: the importance of recognising that 'the people who suffer the consequences of decisions are the best judges of their impacts, and therefore their fairness'.

Considering this statement in relation to the four described resilience themes suggested a larger operating paradigm existed, but one I struggled initially to understand. Renn's statement suggests that the perceived *fairness of decisions* could be a two-way process through which interview participants interact with their world, and this perception of fairness (not some absolute and therefore measurable aspect) could be a fundamental operating principle that contributed to individual's resilience. When interviewees pointed out that there were others in a worse situation than themselves, there was an implied sense of inclusion and belonging: life might not be fair, and some individuals were dealt better hands than others, but where individual choice was available, making a *fair* decision was an indicator of someone's character, and as such, significant.

When talking about fairness, Renn and Webler (2011) describe the need to distinguish between *substantive* and *procedural* fairness: an activity or decision is judged fair if the outcome does not privilege some at the expense of others (Substantive Fairness) and if the decision-making process provides sufficient possibilities for all affected parties to be heard and represented (procedural fairness). Modern democratic societies with pluralistic value systems tend to emphasise procedural justice over substantive fairness, and consensus has emerged in democratic systems to apply the egalitarian principle to procedural equity, whereby providing equal opportunities for all parties involved to influence the decision-making process is seen as the universal yardstick for evaluating procedural fairness. An agreement that procedural fairness is confined to equal representation of each affected party has gained wide acceptance (Renn et al. 1996).

> 'But what does an individual think when they perceive that such a universally accepted principle is not delivering Substantive Fairness for their situation, or not even being applied? How does it impact on their perception of agency in the broader system?'

The concept of *fairness* is well established in formal considerations of justice, and Tost and Lind (2010, p. 21) point out that injustice can inspire action for change and lead individuals to seek power for prosocial ends, particularly when 'negative interpersonal treatment or the violation of moral mandates, activate [an individual's] alarm phase'. However, for individuals who do form a judgement of unfairness, 'positive experiences, such as unexpectedly considerate and respectful interpersonal treatment or the support and protection of moral mandates, are unlikely to activate [the] alarm phase'. The concepts of fairness and justice are not necessarily interchangeable though, illustrated in the final vignette.

Vignette 9.1: Tragedy and Tenacity

He turned over in the air as he passed me, 40 feet, and landed face down but with his forearms propped to protect his head. I'd tied that knot, and it had slipped off the rod he was pulling up. We'd forgotten the harness, and I said we'd go back, but he'd said, 'it's only a 5-minute job', and up he'd gone. Couldn't get the house on the radio and no phone here, so my daughter raced off in the Toyota. I held him in my arms. He was unconscious, and his leg stuck out at an unnatural angle and so did his arm, but I knew he wouldn't die. Don't ask me why…I just knew.

Years later both the doctor and nurse separately told me that my son should have died. But that day, cradling him in my arms as he lay broken in the dusty shade of the windmill for the hours it took for the ambulance and then the helicopter to arrive, I just knew he wouldn't. Maybe he's too much like me – pig-headed and won't be told what to do, or maybe it wasn't his time.

He's back to help run the place now and would prefer me to clear off and leave him to it; and I will, but not completely, not yet. My dad gave me the reins when I was about his age, but he was there in town to offer timely advice when I needed it, or maybe drop a hint about the blindingly obvious I kept running into. Because you can't put an old head on young shoulders and anyways…why would you want to. We need both. Sometimes it's best to go like a bull at a gate, and sometimes it's better to let the dust settle before you climb out of the truck to open it. The trick is knowing the difference.

Not that I really want to leave. His kids are the fifth generation of us here and you don't easily walk away from that. It'd be nice to live somewhere green though and see more of the world. It'd be nice too to reflect on things, maybe write a book; because when I look back, a lot has happened in this life of mine, and it's not over yet. Maybe I would like it too much and not come back – maybe?

I think it was hardest for my wife: she'd heard the helicopter go over and then half an hour later heard it go back. She hadn't seen him, hadn't held him, but had to sit in the car for the hours we drove to the hospital. I'd held him, and that helped.

> He was back up that windmill his first-time home. Three months in hospital and I turn around and there he is, 40 foot up. I asked if he was trying to get back on the horse, but he said 'no, just seeing if it brought any memories back'.

This vignette brings together the four resilience themes and illustrates an essential element of the concept of fairness. It demonstrates *situational awareness* (when to stay, when to go), the *capacity to plan* (setting it up so he can come and go), the *capacity to adapt* (the accident, the changed situation, the return) and *social connectedness* (to the land, to his family, including recognition of the hardship life imposes on his wife); and it does this within the reality of limited health-care support. But nowhere does he say it wasn't fair. It is what happened, and it must be dealt with, by both father and son. In fact, the father wondered about a mate who fell 10 ft from a windmill and was straight up walking around yet died next day from a suspected blood clot. 'Where's the justice in that?' he asked.

9.1 Fairness and Trust

In our increasingly complex world, individuals are not able to inform themselves about all threats that they face, so trust is employed to manage personal risk through externalised faith, particularly if the individual knowledge about the hazard is low (Wachinger et al. 2013). Trust is used as a shortcut to reduce the necessity of making rational judgments based on knowledge by selecting trustworthy experts whose opinion can be considered as accurate (Siegrist and Cvetkovich 2000). This is particularly important in times of crisis, where trust in authorities is necessary if their advice is to be accepted. But the fundamental affective dimension of trust – that lubricant of social life (Putnam 2000) – involves honesty, integrity, goodwill and lack of vested interest, all of which can be influenced or diminished by an individual's perception of fairness.

There are always some people who gain in the process of resilience-building while others lose, so 'we cannot consider resilience without paying attention to issues of justice and fairness in terms of both the procedures for decision-making and the distribution of burdens and benefit' (Davoudi et al. 2012, p. 306). But at the policy level, if adaptation to environmental change is solely concerned with maintaining future flexibility, then some people, communities or ecosystems may incur a heavy price in the present. While this trade-off between equity in process and equity in outcome are central questions of governance, 'Inclusion of vulnerable sections of society and representation of vulnerable social-ecological systems within decision-making structures is an important and highly under-researched area' (Nelson et al. 2007, p. 410).

The importance of trust rose time and again in interviews and figured strongly in each theme but was particularly evident in the *Social Connectedness* theme, with theory relating to human and social capital. People are more likely to accept the described ebb and flow of burdens and benefits if they trust the individuals and institutions involved, and the best time to build trust is before it is required –

incrementally, and in the day-to-day interactions between people and their institutions. This requires deliberate and sustained investment in all forms of capital:

- Maintenance and enhancement of material capital so the community does not feel physically diminished
- Development of human capital so people have the skills and capabilities to act in new ways
- Attention to bonding social capital so discrete sectors feel identified, secure and supported, but not polarised or in direct competition with other sectors for resources
- Continual renewal of bridging social capital so resources, competencies and connections are effectively shared between heterogeneous groups
- Moral capital, essential if hard decisions are required that will require trust, and for the cultivation of virtue and justice (from Stokols et al. (2013).

9.2 Implications for Policy

It is apparent that most individuals currently involved in Northern Australian agriculture are doing so with their *eyes wide open*[1]; that is, they are aware of their operating environment, and its associated limitations and risks and are happy to continue their involvement. This is an important point to establish with respect to the enthusiasm in the press and political circles for a rapid expansion of Northern agriculture: it would be very easy to believe that Northern Australia's "time has come" if one just read newspapers, whereas contemporary northern farmers have a realistic perspective of the situation. They do not see themselves in some serendipitous alignment of forces outside their control that are about to bestow largesse or a *river of gold* (Holthouse 1967).[2] Participants recognise that they need to keep their eyes open because there is another side to the story of northern development that isn't so sexy, so exciting, and so intoxicating – it's called reality, and it has precedence in history. There are many past grand northern endeavours that have come to nothing, and farmers in the main are aware of them and have thought about what went wrong, and they do try not to repeat past mistakes. Recognition that northern farmers are keeping their *eyes wide open* is an important component of this study, and I can think of no better way of conveying it as an attribute than through the vernacular, which is so often more succinct yet descriptive than formal language.

[1] An Australian colloquialism, meaning one is totally understanding and aware of a situation, including associated problems. However, it can also serve as a warning that all may not be as it seems: there could be something *lurking in the shadows*, figuratively or literally; or there may be a bigger agenda at work that needs to be considered; or someone could be trying to *pull the wool over your eyes* (deceive you).

[2] Title of the 1967 book by Hector Holthouse describing the wild days of the 1870s Palmer River gold rush on Cape York and a term that has entered the regional vernacular to describe coming into unforeseen wealth.

Continuing with the vernacular, it is also apparent that many farmers will continue to be involved in agriculture *come Hell or high water*[3] – they are excited by the future, and they see opportunity there. While not a special breed, these farmers do have the advantage of experience (their situational awareness); they can plan for contingencies and can adapt their plans as either opportunities or the unexpected arise; and they are sufficiently connected to their social and physical environments to draw appropriate support when required.

Farming carries risk beyond the control of individual enterprises or even entire industries, but making calculated business decisions in the face of uncertainty has not been an historic impediment to farmers *having a go*. They understand their world, and they make it work. There are many risks outside their control, but they can develop strategies to buffer these. Their real risk comes from things outside their knowledge, experience and intuition. But this is no different for most people to a greater or lesser extent; for example, the developed world's communities are being challenged by the robotisation of many jobs as new technologies change the way our world operates.[4]

However, not everyone has an overt external intention applied to grow their world in the way that an expansion of Northern Australian agriculture is being currently promoted. In large metropolises, growth of one industry may have little effect on underlying community values, because subsequent population changes may be small compared to the total population. However, in small communities, rapid growth in particular sectors may start an endogenous cycle of changing values which define future economic trajectories (Esparon et al. 2018). The White Paper on Developing Northern Australia has growth as an explicit policy intention, and what communities decide when they make policy is *meaning*, not *matter*, and science alone cannot settle these questions of meaning (Stone 2012).

While most farmers see opportunity in an expanding agricultural sector and are generally supportive of the initiative, they are in the main looking at the *matter*. There is a responsibility on those making the policy to think beyond today and guard against perverse outcomes for affected participants and to ensure the community is appropriately involved in and truly understands the development and expression of this *meaning*, because the consequences of actions and measures 'must be considered within the much broader social and environmental context; trade-offs and the potential for negative outcomes over space and time must be recognised' (Eriksen et al. 2011, p. 17).

The affected community in this case is all of Australia, and Australia operating within a global context. The small number of farmers who comprise the tiny part of the 5% of Australians currently living in the north seems insignificant with respect

[3]Another colloquialism, approaching biblical, which implies 'I will succeed in this situation **come what may'**. It conveys an attitude of determination, of resoluteness, of a willingness to endure whatever trials and tribulations are encountered along the way without losing sight of the goal.

[4]Frey and Osborne (2013) of the Oxford Martin Programme on the Impacts of Future Technology at the University of Oxford looked at 702 types of work and ranked them according to how easy it would be to automate them. They found that just under half of all jobs in the USA could feasibly be done by machines within two decades.

to the current world population; but these few farmers are the ones who have the knowledge, skills and ability to make their present situation work; and this experience will be critical to the success of any agricultural expansion. The likelihood of achieving a sustainable outcome will be enhanced if measures are implemented that enable this local capacity to be both recognised and utilised, that the voices of vulnerable groups are heard in decision-making processes that affect their interests and that these interests are not diminished in the face of strong or vocal lobbies. Such an approach could also assist in avoiding some of the problems of sustainable development described by Brown (2011), particularly that the concept can be deliberately vague and slippery, making it difficult to operationalise.

Recent years have seen a 'dramatic increase in urbanites' interest in local food' (Cleveland et al. 2016, p. 99), exemplified by the success of farmer's markets, community-supported agriculture, local distribution hubs and lifestyle television that shows such as SBS Australia's *Gourmet Farmer.*[5] Accompanying this has been increasing demands for health, animal welfare and sustainability, and there is a plethora of ways that agriculture has sought to demonstrate that it can meet consumer expectations, from labelling and sustainability certification through to traceability and regional landscape agreements. Yet consumers' level of trust in agriculture has been dropping. This trend parallels the post-productivist agricultural trends in many parts of the developed world, with an inclination for people with pro-environment views to support agriculture, but only 'because they view it as a better option than development, not because they value it per se' (Cleveland et al. 2016, p. 99). Cleveland et al. also point out that 'living in a rural area, growing up or living on a farm, and having social contact with farmers are associated with support for farmers and trust in their ability to farm in environmentally friendly ways' (p. 90), but most Australians do not live in or adjacent to a rural area. The vast expanses and low resident population of Northern Australia diminish post-productive agricultural influences in all but peri-urban areas and the more settled districts of Queensland's Wet Tropics and Atherton Tablelands, and it is conceptually difficult for many urban Australians to comprehend the reality of northern agriculture, particularly when their exposure is more likely to be through adventure travel shows that often emphasise and mythologise the frontier nature of the north.

The challenge for Australian society is to bridge this cultural divide, to build understanding of and respect for the perspectives of various community sectors and to develop trust in each other's ability to make the right decisions, because ignorance[6] exists on both sides, particularly a lack of understanding of what the other party wants. The existing divide should not be interpreted as disempowerment of northern farmers; rather it is a constraint on Australians being able to collectively make informed decisions.

---

[5] SBS television, now in its fourth series www.matthewevans.net.au/what/gourmet-farmer. An interesting facet of this programme is its emphasis on networks and mates.

[6] Ignorance is used here as a neutral term, to indicate the lack of understanding or awareness.

9.3 The Benefit of Including Fairness in Policy

Most farmers are farming because they want to farm, and in many respects, they are more than capable of continuing to do so. To some degree, all society should do is let them get on with it without undue or unreasonable constraint, other than the obvious ones of abiding by the law and respecting their responsibilities as land managers for good environmental stewardship. But when any group of people (not just farmers) abide by the rules of a society and contribute to its wellbeing, society has a responsibility to recognise and value their role, on some occasions support it, and (within reason) ensure that they are neither forced out of their role nor forced to continue in it against their will.

There are many options for northern development. Every scenario has its champions and its critics; but in making these big policy decisions, providing procedural fairness for the decision process will not be enough: neither for the individuals who will be most affected (the residents), nor for the landscape and industries in which they will operate. The endemic knowledge of the landscape and how its rural industries operate will need to be considered and included as a deliberate act, for example:

1. If a non-resident corporation could purchase and control both the management and long-term destiny of vast tracts of Northern Australia without consideration of local community views and aspirations, would this be considered fair and accepted on the regional stage?
2. If the Australian Government were to allow a foreign entity to invest in Northern Australian property, industry or infrastructure without assisting them to properly understand the unique operational parameters of the region (that is, knowledge that a local would be reasonably assumed to have), would this be considered fair on the international stage, or would the principle of caveat emptor[7] be accepted and no grudge carried forward?
3. If a development or legislative change compromised future generation's ability to enjoy lifestyle and values available today, how would history judge us? Specifically, consider the free-holding of leasehold land which delivers a one-off return to the state, but in so-doing transfers the opportunity for future higher value transactions into private hands.

The four themes identified in this book are personal attributes, but not stand-alone ones. Fairness is what relates the themes back to the broader agency; and this dual focus on agency and structure, the individual and wider society, is a distinctive feature of social research (Oliver 2012). Fairness is a substantial driver of public expenditure to improve housing for remote Indigenous communities, and in the continual improvement of regional health services. Fairness is the micro component of the macro issue of unintended consequences from policy: individual farmers are

[7] The principle that the buyer alone is responsible for checking the quality and suitability of goods before a purchase is made.

distant from the policy-makers but suffer the consequences. Fairness is different from equity and power and it is not a synonym for either dis-empowerment or entitlement, because fairness can also be an enabler of the resilience individuals derive from the implementation of their personal strategies. Risk acceptability is less about accurate probabilities than it is about subjective values and the perception of distributive justice (Renn 2006).

There are risks and benefits associated with an expansion of Northern Australian agriculture, but these risks and benefits are not mutually exclusive. Public deliberation needs to go beyond a narrow focus on minimising risk to achieve a narrowly defined benefit, because as Gross (2007) explains this will enable the social system to be resilient to ignorance and non-knowledge, and thereby resilient to hazards that are not imaginable because they are beyond the system's horizon of expectations. Public deliberation needs to investigate opportunities to maximise benefits in ways that build local capacities and enhance resilience to risk (Wong 2015), but this deliberation will need assistance in understanding that change is a normal part of the persistence of systems (Lorenz 2013). Policy processes that enable this thinking while preventing the hijacking of decisions by vocal or well-organised lobby groups are required, because including affected publics in decision-making processes will help solve many barriers to change, improve resilience, and ultimately produce better decision outcomes.

Historically, humanity has put great faith in technological innovation to help transform societies and improve the quality of life, assisted by the neo-liberal belief that the free market would deliver innovation and improve the quality of life for everyone. While there are many instances where this experiment has proved disappointing and it is now acknowledged that purely technical or structural solutions will not provide absolute protection against the negative impacts of natural hazards (Schanze et al. 2008), society has not lost confidence in the technological fix. However, while the 'spread of material wealth … is closely tied to the maintenance of the social peace enjoyed by western countries' (Westley et al. 2011, p. 763), economic and population growth have compromised many of the planet's ecosystem services and today's connected global society increasingly rejects continuation of such practices.

Academics, scientists, and policy-makers expect facts and figures to carry an argument but, to the surprise of the proponents, emotion often wins over reason in public debates. Westley et al. state that expert-driven, centralised, and top-down approaches to solving these problems are not nimble enough to effectively address the challenges, which are characterised by high levels of complexity and uncertainty; but neither are traditional, disciplinary-based research approaches. Rather, they suggest that new forms of knowledge integration and generation capable of including diverse ideas and viewpoints are required to bring research and action into closer proximity. Brown (2016) suggests a resilience approach might provide the analytical insights into such maladaptation and help formulate policies which avoid them. However, as Berkes and Ross (2013, p. 17) point out

> 'An integrated concept of community resilience is not only about theory; it is equally about practice: How can adaptive capacity, self-organization and agency be supported and fostered through processes such as community development and community-based planning?'

Both individual and society matter in the landscapes of Northern Australia, as they exist on different ontological levels, possess distinct properties and cannot be subsumed into one another (Oliver 2012, p. 381). Farmers have the demonstrated capacity to survive and prosper in their landscape, and development of an integrated model of community resilience will require the overt inclusion of existing farmers and their individual perspectives. While not suggesting that farmers' current agricultural practices represent the pinnacle of sustainability (there are many examples of mistakes made) the aspiration of most farmers is to maintain their resource, their landscape, and their communities. Appreciation and inclusion of this knowledge will benefit and enrich policy decision-making. That this is an entirely realistic aspiration is demonstrated through the improved regional health service achieved when local systems are understood, and knowledgeable and experienced locals are able to contribute to planning and service delivery in a flexible, context-derived manner.

The understanding of resilience as a lived individual actor experience that has become evident through this study has also enabled recognition of the possibilities for the perception of fairness to bring together all the factors at play. In the same way that 'resilience cannot be directly observed or measured – hence they refer to it as a metaphor' (Brown 2016, p. 161), fairness can be a metaphor for ensuring a deliberate process for the engagement, consideration, and inclusion of impacted communities is adopted for policy development, particularly for policy that bonds communities and cultures within their environment.

9.4 The Need to Extend Fairness Through Governance

It has been described how a political decision based on one television documentary had profound and far-reaching impacts on the Northern Australian beef industry, but if we were to look at the same situation from the perspective of an inner-city Australian confronted by abhorrent images of animal cruelty – the immediate cessation of live cattle export was arguably an acceptable response to a situation they could not condone but had no power to influence. It was an emotional response, not a reasoned response. While the event is indicative of the power of a small but motivated and focused sectoral group to influence voters, and thereby direct policy decisions, it is equally important to consider the context of contemporary aspirations for Northern Australian agricultural development:

- It's a long way from where most Australians live;
- Very few people live there, and even fewer are farmers;
- There are limited opportunities for social engagement or sharing of perspectives between Australia's urban majority and its northern residents;
- The nightly news continually tells everyone that the planet is in crisis.

So, from the perspective of a majority of Australians, why not lock up a remote but identifiable piece of 'pristine' country before it is destroyed?

Individual farmers have no control over livestock slaughter in another country, but they do directly bear the consequences of related policy decisions. Whose responsibility is it to ensure that divergent perspectives are properly communicated between and considered by all parties so that the full ramifications of any decision are properly understood before the decision is made?

- Is it the elected governments', usually lacking an on-ground community presence or perspective, and ever-mindful of the 24-h news cycle;
- Is it the various agricultural industry representative organisations, invariably under-resourced and trying to respond simultaneously to a myriad of issues;
- Is it the individual farmers', remote and working hard, often limited in their ability to spend the required time understanding the entirety of an issue?

Verweij et al. (2006, p. 839) describe the usual outcome of such complex and competing multi-player situations as one of extremes, with 'an unresponsive monologue' at one end and a 'shouting match amongst the deaf' at the other; whereas what is required is 'a vibrant multivocality in which each voice formulates its view as persuasively as possible, sensitive to the knowledge that others are likely to disagree, and acknowledging a responsibility to listen to what the others are saying'. Verweij et al. recognise this approach as 'clumsy', yet believe society must strive for such a vocal multivocality 'if we value democracy'; though 'getting there and staying there is … not easy'.

9.5 Opportunities to Improve Fairness Through Governance

The investment case for conserving healthy and productive landscapes will only grow stronger. Australia's established international brand as a supplier of clean, healthy and sustainably grown food and fibre, and of nature-based tourism, positions it well to attract private sector investment, and there more and more examples to disprove the perception that sustainability does not produce finance returns.

There is a crucial difference between contemporary aspirations for Northern Australian agriculture growth and those of previous eras – the existence of Regional Natural Resource Management Groups. There are 56 regional NRM organisations across Australia that act as delivery agents under the National Landcare Programme (Australian Government 2016), and they are the only community-based organisations with the sole purpose of working with all stakeholders to address large and complex natural resource issues at the landscape level; building collaboration, gathering and sharing information, and brokering funding for on-ground work. Their work is done in a planned and integrated manner to deliver effective and efficient government programmes that leverage investment from a wide range of community, industry and business partners.

The membership base of Regional NRM Groups includes local government, Traditional Owners, Lsrandcare and Catchment groups, agricultural industries, conservation groups, and other land managers; and provides a platform for enduring links into communities through community-selected governing boards typically supported by organisational capabilities built over the past 20 years. This broad membership has been shown capable of achieving a balanced, middle ground approach to identifying and implementing desired social, economic and environmental outcomes, while providing value for Government investment without the overheads of departmental delivery.

Since their inception Regional NRM Groups have built partnerships with industry, community, government and other sectors to address complex NRM challenges that span large areas and impact on economic opportunities, social wellbeing, and environmental assets; such as Great Barrier Reef water quality through a *Reef Alliance*[8]. Increasingly, people understand governance to provide a wider set of processes for bargaining and negotiation among differing interests in society that can deliver public and private good outcomes (Dorcey 1986), and Regional NRM Groups are uniquely placed to achieve improved governance of Australia's natural, agricultural and community assets through provision of a trusted third-party perspective and authentication between Australia's dispersed regional communities and its predominately urban population, one that could provide opportunities for information sharing and understanding. Literature recognises the imperative for building and engaging social and human capital as a precursor to effecting the changes that will lead to improved resource condition, and building adaptive capacity that will enable communities to respond more effectively to future sustainability challenges (Argent 2011; Curtis et al. 2014;Dale 2013; Dale et al. 2013; ISSC/UNESCO 2013; Larson 2006; Larson and Brake 2011; Sayer et al. 2015; Smajgl and Larson 2007; Taylor 2010; Vella and Dale 2013; Verweij et al. 2006).

Landscape-scale planning and management brings together diverse land managers, sectors and stakeholders to coordinate planning and implementation across an entire landscape. It seeks to coordinate vertically and integrate horizontally, helping connect, align and coordinate across scales. This offers the potential to reconcile competing objectives at different scales. Landscape-scale planning is important because significant natural resource and triple-bottom line outcomes cannot be achieved through ad hoc property or local-level activity.

Given the complexity of landscape processes and the multiple decision-makers who influence landscape outcomes, Regional NRM Groups stand out as having the demonstrated requisite to build and maintain both social capital in rural areas, and to embed NRM planning that supports transitions and manages trade-offs. This provides strong rationale for governments to support the basic infrastructure required to sustain Australia's Regional NRM bodies as they explore better ways of ensuring regional communities are represented in legitimate, effective and fair governance of Australia's landscapes.

[8]The Reef Alliance brings together industry and regional NRM bodies and conservation sector with a common goal of assisting to secure the future health of the Great Barrier Reef and supporting engaged and prosperous communities.

References

Argent, N. (2011). Trouble in paradise? Governing Australia's multifunctional rural landscapes. *Australian Geographer, 42*(2), 183–205. https://doi.org/10.1080/00049182.2011.572824.

Australian Government. (2016). National landcare programme. Retrieved from http://www.nrm.gov.au/regional

Berkes, F., & Ross, H. (2013). Community resilience: Toward an integrated approach. *Society & Natural Resources, 26*(1), 5–20. https://doi.org/10.1080/08941920.2012.736605.

Brown, K. (2011). Sustainable adaptation: An oxymoron? *Climate and Development, 3*(1), 21–31. https://doi.org/10.3763/cdev.2010.0062.

Brown, K. (2016). *Resilience, development and global change*. London: Routledge.

Cleveland, D. A., Copeland, L., Glasgow, G., McGinnis, M. V., & Smith, E. R. A. N. (2016). The influence of environmentalism on attitudes toward local agriculture and urban expansion. *Society & Natural Resources, 29*(1), 88–103. https://doi.org/10.1080/08941920.2015.1043081.

Curtis, A., Ross, H., Marshall, G. R., Baldwin, C., Cavaye, J., Freeman, C., et al. (2014). The great experiment with devolved NRM governance: Lessons from community engagement in Australia and New Zealand since the 1980s. *Australasian Journal of Environmental Management, 21*(2), 175–199. https://doi.org/10.1080/14486563.2014.935747.

Dale, A. (2013). *Governance challenges for northern Australia.* (978-0-9875922-1-7). Retrieved from http://www.jcu.edu.au/cairnsinstitute/public/groups/everyone/documents/working_paper/jcu_128813.pdf

Dale, A., McKee, J., Vella, K., & Potts, R. (2013). Carbon, biodiversity and regional natural resource planning: Towards high impact next generation plans. *Australian Planner, 509*, 1–12. https://doi.org/10.1080/07293682.2013.764908.

Davoudi, S., Shaw, K., Haider, L. J., Quinlan, A. E., Peterson, G. D., Wilkinson, C., et al. (2012). Resilience: A bridging concept or a dead end? "Reframing" resilience: Challenges for planning theory and practice interacting traps: Resilience assessment of a pasture management system in northern Afghanistan urban resilience: What does it mean in planning practice? Resilience as a useful concept for climate change adaptation? The politics of resilience for planning: A cautionary note. *Planning Theory & Practice, 13*(2), 299–333. https://doi.org/10.1080/14649357.2012.677124.

Dillon, B., & Bourke, J. (2016). *The six signature traits of inclusive leadership: Thriving in a diverse new world.* Retrieved from http://www2.deloitte.com/au/en/pages/human-capital/articles/six-signature-traits-inclusive-leadership.html?Id=au:2sm:3li:4dcom_share:5awa:6dcom:human_capital

Dorcey, A. H. J. (1986). *Bargaining in the governance of Pacific coastal resources : Research and reform/Anthony H.J. Dorcey*. Vancouver: Westwater Research Centre, Faculty of Graduate Studies, University of British Columbia.

Eriksen, S., Aldunce, P., Bahinipati, C. S., Martins, R. D., Molefe, J. I., Nhemachena, C., et al. (2011). When not every response to climate change is a good one: Identifying principles for sustainable adaptation. *Climate and Development, 3*(1), 7–20. https://doi.org/10.3763/cdev.2010.0060.

Esparon, M., Farr, M., Larson, S., & Stoeckl, N. (2018). Social values and growth and their implications for ecosystem services in the long-run. *Australasian Journal of Regional Studies, 24*(3), 327–346.

Frey, C. B., & Osborne, M. (2013). *The future of employment: How susceptible are jobs to computerisation?* Retrieved from Oxford Martin Programme on Technology and Employment: http://www.oxfordmartin.ox.ac.uk/downloads/academic/future-of-employment.pdf.

Gross, M. (2007). The unknown in process: Dynamic connections of ignorance, non-knowledge and related concepts. *Current Sociology, 55*(5), 742–759. https://doi.org/10.1177/0011392107079928.

Holthouse, H. (1967). *River of gold : The story of the Palmer River gold rush*. Sydney: Angus and Robertson.

ISSC/UNESCO. (2013). *Earth system governanc*. Retrieved from OECD Publishing and UNESCO Publishing

Larson, S. (2006). Analysis of the water planning process in the Georgina and Diamantina catchments: An application of the Insitutional Analysis and Development (IAD) framework. *ResearchGate*: CSIRO Sustainable Ecosystems, Townsville and Desert Knowledge CRC, Alice Springs.

Larson, S., & Brake, L. (2011). Natural resources management arrangements in the Lake Eyre Basin: An enabling environment for community engagement? *Rural Society, 21*(1), 32–42.

Lorenz, D. F. (2013). The diversity of resilience: Contributions from a social science perspective. *Natural Hazards, 67*(1), 7–24. https://doi.org/10.1007/s11069-010-9654-y.

Maddi, S. R. (1999). The personality construct of hardiness: I. effects on experiencing, coping, and strain. *Consulting Psychology Journal: Practice and Research, 51*(2), 83–94. https://doi.org/10.1037/1061-4087.51.2.83.

Nelson, D. R., Adger, W. N., & Brown, K. (2007). Adaptation to environmental change: Contributions of a resilience framework. *Annual Review of Environment and Resources, 32*(1), 395–419. https://doi.org/10.1146/annurev.energy.32.051807.090348.

Oliver, C. (2012). Critical realist grounded theory: A new approach for social work research. *British Journal of Social Work, 42*(2), 371–387. https://doi.org/10.1093/bjsw/bcr064.

Putnam, R. D. (2000). *Bowling alone: The collapse and revival of American community*. New York: Simon & Schuster.

Renn, O. (2006). Participatory processes for designing environmental policies. *Land Use Policy, 23*(1), 34–43. https://doi.org/10.1016/j.landusepol.2004.08.005.

Renn, O. (2015, 14 July 2015). [Cairns Institute Workshop on Risk Governance].

Renn, O., & Webler, T. (2011, 8 Mar 2016). [Fairness in Risk-Based Decision-Making: Implications for Citizen Participation].

Renn, O., Webler, T., & Kastenholz, H. (1996). Procedural and substantive fairness in landfill siting: A Swiss case study. *Risk: Health, Safety, and Environment, 7*, 145–168.

Sayer, J., Margules, C., Bohnet, I., Boedhihartono, A., Pierce, R., Dale, A., & Andrews, K. (2015). The role of citizen science in landscape and seascape approaches to integrating conservation and development. *Land, 4*(4), 1200.

Schanze, J., Hutter, G., Penning-Rowsell, E., Nachtnebel, H., Meyer, V., Werritty, A., …, Schildt, A. (2008). Systematisation, evaluation and context conditions of structural and non-structural measures for flood risk reduction. Retrieved from London: www.crue-eranet.ne

Scullion, N. (Producer). (2016, 10 February). Indigenous affairs minister opens up about Closing the Gap. *RNDrive*. Retrieved from www.abc.net.au/radionational/programs/drive/indigenous-affairs-minister-opens-up-about-closing-the-gap/7157660

Siegrist, M., & Cvetkovich, G. (2000). Perception of hazards: The role of social trust and knowledge. *Risk Analysis, 20*(5), 713–720. https://doi.org/10.1111/0272-4332.205064.

Smajgl, A., & Larson, S. (2007). *Sustainable resource use: Institutional dynamics and economics*. Sterling: Earthscan.

Stokols, D., Lejano, R. P., & Hipp, J. (2013). Enhancing the resilience of human–environment systems: A social ecological perspective. *Ecology and Society, 18*(1), 7. https://doi.org/10.5751/ES-05301-180107.

Stone, D. (2012). *Policy paradox : The art of political decision making*. New York: W.W. Norton & Co.

Taylor, B. M. (2010). Between argument and coercion: Social coordination in rural environmental governance. *Journal of Rural Studies, 26*(4), 383–393. https://doi.org/10.1016/j.jrurstud.2010.05.002.

Tost, L. E., & Lind, E. A. (2010). Sounding the alarm: Moving from system justification to system condemnation in the justice judgement process. In M. A. Neale, E. A. Mannix, & E. Mullen

(Eds.), *Research on managing groups and teams, volume 13 : Fairness and groups* (pp. 3–28). Bradford: Emerald Group Publishing Ltd.

Vella, K., & Dale, A. (2013). An approach for adaptive and integrated environmental planning to deal with uncertainty in a great barrier reef catchment. *Australian Planner, 51*(3), 243–259. https://doi.org/10.1080/07293682.2013.837831.

Verweij, M., Douglas, M., Ellis, R., Engel, C., Hendriks, F., Lohmann, S., & Thompson, M. (2006). Clumsy solutions for a complex world: The case of climate change. *Public Administration, 84*(4), 817–843. https://doi.org/10.1111/j.1540-8159.2005.09566.x-i1.

Wachinger, G., Renn, O., Begg, C., & Kuhlicke, C. (2013). The risk perception paradox—Implications for governance and communication of natural hazards. *Risk Analysis, 33*(6), 1049–1065. https://doi.org/10.1111/j.1539-6924.2012.01942.x.

Westley, F., Olsson, P., Folke, C., Homer-Dixon, T., Vredenburg, H., Loorbach, D., et al. (2011). Tipping Toward Sustainability: Emerging Pathways of Transformation. *Ambio, 40*(7), 762–780.

Wisner, B., Blaikie, P., Cannon, T., & Davis, I. (2003). *At risk second edition: Natural hazards, people's vulnerability and disasters*. London: Routledge.

Wong, C. M. L. (2015). The mutable nature of risk and acceptability: A hybrid risk governance framework. *Risk Analysis*, n/a-n/a. https://doi.org/10.1111/risa.12429.

Chapter 10
A Fair Go, a Fair Future

Since it is the writing itself that leads to insight, we must become used to starting off in a state of unknowing and letting the writing lead us to whatever memories we require.
Sheila Bender (2015)

Agriculture will continue to be an important component of North Australia's economy and identity, though its scale, ownership and focus could develop in myriad and possibly unexpected ways. This book has not been an attempt to predict this future.

That Northern Australia will not develop in the same historical manner as the more populated south is a reasonable assumption – it is different geographically, climatically and demographically, and further development will occur in today's globally connected and informed community, very different from the colonial development of Southern Australia. Nevertheless, it remains important to understand and always consider a region's history as a component of development planning, as past and present aspirations combine with contemporary trends and influences to determine the future. This is necessary not to understand agriculture, but to understand the operational context and perspective of farmers involved in Northern Australian agriculture.

That agriculture will continue to be a part of Northern landscapes is also a reasonable assumption – the area is vast and capable of accommodating intensive and extensive agriculture along with tourism and the *light*[1] touch of conservation management.

Keith Noble has contributed more to this chapter.

---

[1] The setting aside of areas for conservation management would be seen by many as a lower impact land use than agriculture, mining or urban development, that is, a *lighter* touch. In some instances though, setting aside areas for conservation management can exacerbate weed and feral animal problems, change fire regimes or deliver unintended outcomes. All land requires management and commensurate resource allocation if anticipated outcomes are to be achieved.

K. Noble et al., *Agriculture and Resilience in Australia's North*,
https://doi.org/10.1007/978-981-13-8355-7_10

Australian agriculture has a track record of speedy and successful innovation and technology adoption that will continue through genetics, remote sensing and Internet/device connectivity, while automation and customers' ability to trace food from farm-to-fork will continue to influence supply chains (Hajkowicz and Eady 2015).

The third assumption is that North Australia's agricultural future is fundamentally tied to the future of its First People, because as majority landholders and, in many areas the majority of the population, until their interests are optimally accommodated, Northern Australia will never reach its social and economic potential and at worst, could fail to develop socially and economically.

Our era provides an unprecedented opportunity to consider and debate development options before their adoption, should Australians as a community choose to do so. Such a debate should be informed and guided by the lessons of history, along with the contemporary wisdom and experience of others around the world, facilitated by communication technologies and social media enabling whole-of-community participation to collectively agree on a shared future. While fictional expectations drive modern economies – or throw them into crisis when the imagined futures fail to materialise (Beckert 2016) – the big risk in this opportunity, and the intention of this book to address, is that the voices of those with direct experience and practical understanding of the reality of agriculture in the north could be lost, overwhelmed or disregarded in the conversation.

10.1 Resilience and North Australian Agriculture

Change is unpredictable, so communities are unlikely to have full knowledge of the changes to anticipate or the intensity or ultimate impact of those changes (Magis 2010). The increasing interest in and invocation of the notion of resilience within communities to better manage such change has been discussed, and the concept has entered national, regional and local policy discourse. While acknowledging that the resilience concept 'suffers from imprecision of definition and conceptualization, which in turn weakens its purchase as an analytical or explanatory tool' (Martin 2012, p. 26), the observation by Hassink (2010) that the study of regional resilience is fraught with methodological and philosophical difficulties and therefore 'its contribution is relatively limited' (p.45) is at odds with the enduring and expanding cross-professional interest in the concept, ongoing community use of the term and its copious representation in literature as described in Chap. 2.

Resilience's malleability and capacity to bring together multiple disciplines and understandings explains in part its persistence and importance to individual, community and social sustainability. Christopherson et al. (2010) suggest that maybe the attention to resilience is a response to a generalised contemporary sense of uncertainty and insecurity and a search for formulas for adaptation and survival, perhaps because processes associated with globalisation have made places and regions more permeable to the effects of what were once thought to be external processes. People look for a safe (imagined) past where they knew the rules – because they don't know the rules of the future.

Agriculture has proved itself a dynamic and innovative industry capable of adopting new technology and aligning itself with social trends, and society's fears around issues like robotisation of the workforce are viewed as real opportunities for some dispersed and labour-poor northern agricultural sectors, allowing farmers to perform jobs like extensive weed survey or resource monitoring which are currently either unaffordable or onerous on the individual if done manually. While the ability to capitalise on such innovation is hampered by present NBN Internet service delivery – a particular disservice when high-speed Internet would improve planning and adaptation capacity, enable family connectedness and facilitate increased situational awareness – innovations in IoT connectivity combined with edge computing is enabling outcomes today.

However, provision of a fast universal broadband network to all Northern Australians would not be a panacea for all the challenges faced by the region's farmers. The history of northern development described in Chap. 3 illustrates the gradual but inexorable nature of development and the sometimes piecemeal and often politically influenced *official* approach to development and infrastructure provision, which is really what needs to change. Retrospection[2] indicates that narrow-based development manifestos are unlikely to achieve their envisioned success, which should be noted and remembered by policy mandarins and business entrepreneurs.

10.2 Resilience and North Australia

While the future of Northern Australia is rooted in this past, the future will manifest in the complex milieu of a global society, and the social complexity that accompanies a growing world population (described in Chap. 4) allows many possible and contested futures. Prospection[3] is a ubiquitous feature of the human mind (Seligman et al. 2013), and the ability of people to 'pre-experience … the hedonic consequences of events they've never experienced by simulating those events in their mind' (Gilbert and Wilson 2007, p. 1351) is clearly at work in the imagined futures for Northern Australia, and 'these prospects can include not only possibilities that have occurred before but also possibilities that have never occurred – and these new possibilities often play a decisive role in the selection of action' (Seligman et al. 2013, p. 119). While no one group is likely to determine Northern Australia's ultimate destiny, it is of paramount importance to consider and include the imaginings of those already living and working there.

It is important because of the specific regional knowledge and understanding that they hold, and it is important if Australia is to maintain its reputation as a fair and compassionate society. The ability of communities to do so is demonstrated by the experience, outcomes and innovations in diverse sectors, with examples provided in architecture (Chap. 7) and health (Chap. 8). Safe and social living and work envi-

[2] The ability to re-experience the past.

[3] The generation and evaluation of mental representations of possible futures.

ronments are achieved through acknowledging and including local knowledge in architecture, and improved health service delivery to remote areas achieved via bottom-up changes developed with the practical experience of real-world practitioners living in and with the affected community. And it is in Northern Australia that so many of Australia's First People have maintained and continued to adapt their time-proven and fine-grained understanding of how humans can and do live in this place.

The question of a community's resilience is an old and enduring one, and individuals understand that their own ability to successfully respond to change is enmeshed with their collective resilience, which in turn is a composite of the numerous heterogeneous entities and individuals that compose the community, and their interaction. Through their own words and describing their individual lived experience, interviewed farmers understand resilience as not an immutable characteristic that an individual or a community has or does not have but as a process that emerges from malleable resources. What then are the opportunities for resilience thinking in development of policy?

10.3 Policy and Resilience

An understanding of how farmers think about and interact with their situation and of the interrelationships around this thinking was sought, and it was important this knowledge was sourced directly – the individual's lived experience – rather than filtered through a pre-existing theoretical construct, because as suggested by Lorenz (2013, p. 10):

> 'social resilience does not mean that the system changes as fast as possible or perpetuates its structure under any circumstances, but that the new structure in the case of change involves sustainable variances that enable the system to persist into the future under any given terms'.

This study of the context, personal strategies, perspectives and operating environment of individuals within Northern Australian agriculture (now and in the past) conducted through semi-structured interview with individual farmers has identified and assisted understanding of the factors and strategies that contribute to or enhance an individual's chance of achieving what they perceive as successful outcomes. The inclusion of perspectives from other disciplines of great importance to Northern Australian development, health and architecture, by experienced co-authors, illustrates that the consideration, understanding and inclusion of local context and knowledge improve their resilience planning too. So, can this approach be adopted to improve planning and policy outcomes, particularly in reducing the risk of new industry development or expansion?

The study was necessarily broad but draws on a breadth of relevant author experience, including farmer; architect; doctor; marketer; industry advocate; active NRM and community participants; planning, health care and built form research; and as deliberate northern residents who have worked with and demonstrated interest in the well-being of the people of Northern Australia for a long time. Even so, probably only a superficial understanding of why farmers stuck to such a hard game – what made them resilient – has been achieved, but it does reveal taken-for-granted meanings and portray the fullness of experience to help explain why resilience is important. While every farmer is different from every other farmer, a universal farmer's resilience accord can be seen operating in Northern Australia.

Failure to include this specific regional knowledge, experience and understanding in past policy is seen to contribute to not achieving desired outcomes or, in some cases, not even starting with desirable goals. This deficiency has been exacerbated when decisions are made from major metropolitan centres geographically and culturally removed from the region. Further complicating the situation, the physical and cultural diversity of Northern Australia mitigates against a *one size fits all* policy response and suggests that a diverse and rich array of bespoke policy responses are required under an overarching but agreed intent. The other real risk of not including regional understanding in policy is that processes become vulnerable to hijack by sectoral lobbies or those with strongly held views. The future of a region and its residents should not be jeopardised by policy failure to properly understand or, even worse, to operate on assumptions. This book demonstrates that it is possible to understand the context of those already involved in any activity rather than assuming an understanding exists, but to do so requires deliberate effort.

10.4 An Imagined Future

This book is about achieving a meaningful understanding and appreciation of individual resilience processes and understanding how this can contribute to northern development. Direct involvement of stakeholders was important because they possess knowledge of the local environment and their management strategies have been developed and adapted, often over generations, and then shared and readapted between families and across industries. Historically, such forms of knowledge often fell outside formal science frameworks even though they were demonstrably effective and useful when applied in the local context by experienced practitioners.

But change is afoot: in the same way that the experience and knowledge of health-care professionals is delivering real improvements and cost-savings in service delivery that contribute to individual and community resilience, community-derived landscape management research methodologies, such as the Terrain NRM and NQ Dry Tropic Major Integrated Projects (MIP) to improve Great Barrier Reef water quality (described in Chap. 6), are enabling real-world knowledge and experience informed by science to directly influence policy decisions.

Another key conclusion in the 2016 Queensland Water Quality Science Taskforce report (that resulted in MIP funding) was the need to consider incentives and market mechanisms 'to complement and integrate with regulation extension and education'. Both public and private investment can support sustainable land use directly through existing, maturing and innovative new financing opportunities (or 'incentives'), as well as indirectly, via the private sector's increasing need to price in and manage supply-chain risk ('disincentives') through diversifying away from business models that degrade natural capital.

The investment case for conserving healthy and productive landscapes is been getting stronger as Australia's established international brand as a supplier of clean, healthy and sustainably grown food and fibre, and of nature-based tourism, positions it well to attract private sector investment. Financial opportunities include low-interest loans, green bonds, impact investing in real assets, environmental markets, environmental impact bonds (EIBs) and philanthropic giving. Investors are requiring new finance criteria be met to mitigate the downside legal, market and reputational risks associated with goods and services that degrade environmental assets.

Such market-based approaches work effectively around the world in the provision of environmental services, for example, sequestration of carbon, reduction of sulphur oxides and creation of water quality credits; and the Reef Credit Scheme[4] developed by Terrain NRM in association with GreenCollar[5] adopts a similar approach in the Great Barrier Reef catchments, allowing landowners to generate and sell Reef Credits that result from activities that reduce sediment, pesticide and nutrient losses and may be sold to a range of buyers such as government, corporate, industrial or philanthropic entities. In 2019, Terrain NRM also received state government funding to further explore the concept of Cassowary Credits as a mechanism to improve land managers' ability to protect this iconic but endangered species.

Through a combination of innovation and trusted partnerships, community-based NRM organisations now occupy a unique position in Australia's landscape management, with established relationships and respected communication processes that span the productive, conservation and community sectors. Their ability to be the vehicle for exploring fairness as a contributor to regional policy development and delivery is worthy of support.

An exciting opportunity lies in further exploration of Gammage's (2011) postulate that pre-European Aboriginals were farmers without fences, who shaped the landscape in ways that facilitated plant and animal harvest adapted to the physical constraints of regional soil and climate; and it is possible such an approach might rediscover fundamentally different ways of farming Northern Australia that might help bridge the sectoral disconnect between aspirations for a productive landscape and for the preservation of ecological values. Certainly, the size of Northern Australia's Indigenous population, the importance of Indigenous organisations and extent of Indigenous interests in land, water, sea, natural resources and other intangible assets in Northern Australia, means that Northern Australia's First People are again paramount stakeholders in the Northern Australian economy. This strongly suggests that unless Indigenous interests in the Northern Australian economy are

---

[4] www.reefcredit.org.

[5] GreenCollar is one of Australia's largest environmental markets investors and project developers, committed to working with landholders to design appropriate individual projects which complement existing operations while achieving positive commercial and environmental outcomes.

optimally activated, Northern Australia will at best never reach its social and economic potential and at worst fail to develop socially and economically.

However, the real risks for both Northern Australia's First People and farmers alike, though particularly for an agricultural expansion, are mostly beyond individual's personal resilience processes and strategies. These are the global perspectives and influences of an urban population with diminishing connection to and understanding of land management and particularly agricultural land management in a Northern Australian context. While individuals cannot hope to control these influences, they can appeal to the broader community's sense of fairness to increase the likelihood of their endemic knowledge being valued and included, rather than overlooked.

The term resilience crosses many academic and professional disciplines and has widespread use in the popular lexicon. This persistence of the word and its ultimate flexibility make resilience suitable as a central policy intent. Brennan (2008, p. 61) tells us that 'Resilient communities exhibit adaptive capacities, established networks, infrastructures, and alliances that allow the community to plan for its needs and build on its strengths to achieve desired goals'. By describing resilience from the *bottom up* and through the actual words of those involved, it has been shown that these same interactions are at play at the individual level. This has been established through the words of those involved, so this understanding is not based on an assumption; and it provides an authentic bridge between the academic interpretation of farmers resolve and the farmers themselves.

By establishing this credible association between the academic resilience text and the applied practices of Northern Australian farmers and other community members, the value of relating theory to people *coming through* adversity, both the processes they use and what resilience looks like on the way through, is established. For as Richardson (2002, p. 319) describes

> 'Resilience or energy comes from within the human spirit or collective unconscious of the individual and also from external social, ecological, and spiritual sources of strength. Resiliency and resilience can be seen as simple and practical applications to everyday living'.

This work describes the resilience process in the words of those who have lived it. The research participants have acted without thinking of it as a process with some beginning, middle and end – it has been a part of their life; but they have known they were not alone in the experience. Such experience is a part of every human life, though the *rawness* of the direct interaction with nature that many farmers experience can be diminished when people live further from or are insulated from the cause and effect of seasons and the physicality of life. While everyone's situation is different, there is usually comfort in knowing that there are others *shoulder to shoulder*, sharing the load; and attention to the places people live will also improve their situation. But there is a real risk when this relationship is put at risk through perceptions of *unfairness*. People can feel alienated from their broader society, and as a consequence their valuable knowledge and understanding might be either withheld or excluded from important decision-making processes. Even worse, their wisdom can be openly ignored when freely offered. An appreciation and inclusion of such knowledge is more apparent in a model of resilience developed from the *bottom-up*.

While documenting, considering and understanding these tangible contributors to individual farmer's resilience, the critical importance of *fairness* to farmers became apparent as a connecting concept and translational instrument between their personal worlds and their relationship with the broader community. It is proposed that an understanding and consideration of *fairness* from the perspective of those directly impacted could be an enabler of analytical insights into social ecological systems and therefore should be an essential consideration of policy development relating to northern development.

It is the Thread of Fairness that provides opportunity for building relationships and improving understanding in both directions – between farmers and the broader community – through paying attention to perceptions of fairness. This opportunity became evident through the understanding of resilience specifically as a lived actor experience, and the consideration of fairness is further proposed as a metaphor for ensuring a deliberate process of engagement, consideration and inclusion of impacted communities in policy development.

The concept and reinvention of social ecological systems resilience theory have been driven to an extent by biological scientists who do not always understand the importance of social relationships and capital, these being more usually researched and analysed by social scientists. This book shows that there is a wealth of knowledge and experience at play in farming – some innate, much hard-won – and while a lot of it is directed towards reducing risk in a complex and highly variable physical environment, it is also directed towards reducing risk in the social environment and particularly towards building and maintaining social capital. Social factors such as attachment to place, lifestyle, industry, local knowledge, family, community standing and networks are every bit as important in these social ecological systems.

Success in agriculture is rarely achieved *overnight*, and it is not usually a job someone can just *do* for a couple of years as part of a career smorgasbord. But a core finding of this book is that individual success can be assisted by tying analysis of adaptability, resilience and vulnerability in the biophysical systems to a better understanding of these same features in social and economic systems and by considering this information in situ.

10.5 Concluding Remarks

This book had its genesis in disaster – a personal disaster – and consequently contains aspects of life and understanding derived through having lived the experience. The phenomenological nature of the study uses, in their own words, the lived experience of other individuals as the basis for analysis of resilience strategies. There has been a deliberate attempt to avoid an overly academic reinterpretation of either the strategies or of the language used to describe them. It was the intention that interpretation be informed by resilience literature, rather than filtered through it.

Everyone experiences disasters to greater or lesser degrees – they are a part of life. For some, a particular event might be life-changing, whereas for their immediate neighbour it could be of no consequence, because every person experiences it in the context of not only the event but of their personal situation prior to and subsequent to the event. It then becomes another component of the individual's lived experience which they can consider and draw on in future. How individuals respond to a situation does not make that person better than another, for we are all at the mercy of natural and social events, and that which is the undoing of one can provide opportunity for another. The individual's response is not some badge of honour that can be weighed and compared on a resilience scale.

However, in an awareness of one's situation in life, the ability to plan accordingly and then to be able to review and adapt that plan and at appropriate times call on friends, family or the community for support, this does assist farmers to cope with uncertainty and contributes to their life experience which, as the saying goes, makes them stronger. While this is applicable to people everywhere, it does need to be particularly remembered in situations where the voices of a few could easily go unheard.

Consideration of individual fairness has to be an indispensable component of any policy process, and I am heartened that even as I write this conclusion and on the eve of the announcement of Australia's 2019 federal election, the principal campaign platform of a major party (Shorten 2019) is 'a fair go for all Australians'.

References

Beckert, J. (2016). *Imagined futures: Fictional expectations and capitalist dynamics*. Cambridge, MA: Harvard University Press.

Bender, S. (2015). Personal essayists: What we believe, what we do. Retrieved from www.vineleavesliteraryjournal.com/sowing-the-seeds

Bill Shorten, M. P.. (2019). Labor's fair go action plan. Retrieved from www.alp.org.au/labors-fair-go-action-plan/

Brennan, M. A. (2008). Conceptualizing resiliency: An interactional perspective for community and youth development. *Child Care in Practice, 14*(1), 55–64. https://doi.org/10.1080/13575270701733732.

Christopherson, S., Michie, J., & Tyler, P. (2010). Regional resilience: Theoretical and empirical perspectives. *Cambridge Journal of Regions, Economy and Society, 3*(1), 3–10. https://doi.org/10.1093/cjres/rsq004.

Gammage, B. (2011). *The biggest estate on earth: How aborigines made Australia*. Crows Nest: Allen & Unwin.

Gilbert, D. T., & Wilson, T. D. (2007). Prospection: Experiencing the future. *Science, 317*(5843), 1351–1354.

Hajkowicz, S., & Eady, S. (2015). *Megatrends impacting Australian agriculture over the coming twenty years*. Retrieved from [online]:

Hassink, R. (2010). Regional resilience: A promising concept to explain differences in regional economic adaptability? *Cambridge Journal of Regions, Economy and Society, 3*(1), 45–58. https://doi.org/10.1093/cjres/rsp033.

Lorenz, D. F. (2013). The diversity of resilience: Contributions from a social science perspective. *Natural Hazards, 67*(1), 7–24. https://doi.org/10.1007/s11069-010-9654-y.

Magis, K. (2010). Community resilience: An Indicator of social sustainability. *Society & Natural Resources, 23*(5), 401–416. https://doi.org/10.1080/08941920903305674.

Martin, R. (2012). Regional economic resilience, hysteresis and recessionary shocks. *Journal of Economic Geography, 12*(1), 1–32. https://doi.org/10.1093/jeg/lbr019.

Richardson, G. E. (2002). The metatheory of resilience and resiliency. *Journal of Clinical Psychology, 58*(3), 307–321. https://doi.org/10.1002/jclp.10020.

Seligman, M. E. P., Railton, P., Baumeister, R. F., & Sripada, C. (2013). Navigating into the future or driven by the past. *Perspectives on Psychological Science, 8*(2), 119–141. https://doi.org/10.1177/1745691612474317.